QUENCE
NTS
LAUNCH
CATIONS
FROM
VERAL

2 APPROXIMATELY 26 MINUTES AFTER
~~UP TO F SPACECRAFT~~
~~TO T TRANSFER ORBIT~~

TRANSFER
ORBIT

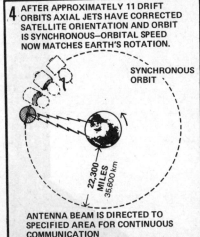

4 AFTER APPROXIMATELY 11 DRIFT
ORBITS AXIAL JETS HAVE CORRECTED
SATELLITE ORIENTATION AND ORBIT
IS SYNCHRONOUS—ORBITAL SPEED
NOW MATCHES EARTH'S ROTATION.

SYNCHRONOUS
ORBIT

22,300
MILES
35,600 km

ANTENNA BEAM IS DIRECTED TO
SPECIFIED AREA FOR CONTINUOUS
COMMUNICATION

Diagram: Hughes
Aircraft Company

The Observer's Pocket Series

UNMANNED SPACEFLIGHT

The Observer Books

A POCKET REFERENCE SERIES COVERING
A WIDE RANGE OF SUBJECTS

The Observer's Book of

UNMANNED SPACEFLIGHT

REGINALD TURNILL

WITH 14 COLOUR AND OVER 50
BLACK AND WHITE ILLUSTRATIONS

FREDERICK WARNE & CO LTD
FREDERICK WARNE & CO INC
LONDON : NEW YORK

LIBRARY OF CONGRESS CATALOG
CARD NO. 73-80242

ISBN 0 7232 1522 7

Printed in Great Britain at
The Pitman Press
Bath

1338.1073

CONTENTS

5

LIST OF COLOUR PLATES

ACKNOWLEDGMENTS

The author and publishers are grateful to Dr D.G. King-Hele for permission to reproduce the latest census of space vehicles from the invaluable Table of Earth Satellites compiled by the Royal Aircraft Establishment at Farnborough, England; and to Dr C.S. Sheldon for similar permission to draw on information in his comparative survey of US and Soviet space progress for the US Library of Congress. Information and photographs have been generously provided, as always, by NASA (the US National Aeronautics and Space Administration), from their centres at Washington, Goddard, Houston, Cape Kennedy, Pasadena, and elsewhere; by TRW, Martin Marietta, McDonnell Douglas, Hughes, and other US aerospace firms; by the US Geological Survey; and by 'Aviation Week & Space Technology'. ESRO and ELDO in Europe; by Hawker Siddeleys, the British Aircraft Corporation and Marconi in Britain. Soviet pictures come from Novosti, Tass and Graham Turnill; the Planetary Flights log is from US sources; research checking and metric conversion by Margaret Turnill.

ABBREVIATIONS

In most cases abbreviations are explained when they first appear in the text. AU stands for Astronomical Unit; dia. for diameter; incl. for inclination; L. for launch; mi for miles; (U) for unnamed.

INTRODUCTION

While US astronauts and Soviet cosmonauts have explored the moon, and learned to live and work for long periods in earth orbit, unmanned spacecraft have been steadily opening up the way to the planets. This companion volume to the *Observer's Book of Manned Spaceflight* seeks to cover this equally exciting and rapidly developing aspect of man's restless, questing spirit, and at the same time to explain *why*, as well as *how* it is being done. It concentrates on the major planetary programmes. The remarkable discoveries made by both US and Soviet scientists with their Mars and Venus programmes, as a result of a decade of perseverance despite many setbacks, are set out first in log form, for quick and easy reference, and then in detailed entries, in alphabetical order under the country concerned. The determined search for life on Mars, now being pressed forward by both Russia and America, is matched by the exciting US PIONEER spacecraft which have penetrated the asteroid belts to seek for us the first pictures of Jupiter and Saturn. Brief, up-to-date sketches of the planets, about which unmanned spaceflight has brought us more information in a decade than man was able to obtain through past centuries of study, are also included.

Because it is now so long, and most of it is covered in *Observer's Manned Spaceflight*, no log is given of flights to the moon; but, of course, the unmanned missions which enabled the Americans to send men to the moon, and which Russia is using for robot exploration, are described under the project names.

8

And short though this book is, it aims to present the clearest, and the most detailed and exciting analysis yet published of the relentless battle between East and West for technological mastery of this new frontier. Soviet military activity is dealt with under COSMOS; the parallel American activities under MILITARY SATELLITES.

It may be that the many people who question whether space exploration is worth while, may find some convincing answers under such headings as NIMBUS, TIROS and ERTS, which are studying the earth's resources and possible ways of harnessing the energy of the sun. Such sceptics will certainly be astonished to learn that Russia's METEOR navigation satellites have enabled their merchant ships to cut their average journey times by 10%. Then there are the developing networks of communication satellites, MOLNIYA, INTELSAT, INTERCOSMOS, which are rapidly changing the lives of all of us.

The limited space available in the *Observer's* books makes it impossible to cover fully all the 120-odd spacecraft projects launched by the 6 countries which have developed orbital launch capability since 1957; but an effort has been made to include some reference, however brief, to most of them.

As in the case of *Observer's Manned Spaceflight*, this volume is designed to be read for interest and also to be used for quick reference. From a reference point of view, it is useful to remember that the Soviet system of using separate project names, i.e. Luna, Mars and Venus, for their lunar and planetary programmes, makes use of the Soviet section relatively easy. In the case of the United States, it should be remembered that Pioneer spacecraft have been sent to all 3 planets; Mars, Venus and Mercury are also being tackled by Projects Mariner and Viking.

SPACE LOGS

Satellites and Space Vehicles 1957–1973

Year of launch

Country of origin	1957–1968	1969	1970	1971	1972	1973	Total national launches 1957–1973
USSR	314	68	79	81	70	83	695
USA	432	33	23	25	24	21	558
France	4	0	1	1	0	0	6
Japan	—	—	1	2	1	0	4
China	—	—	1	1	0	0	2
Britain	—	—	—	1	0	0	1
US/Intelsat	6	3	3	2	2	1	17
USSR/Intercosmos	—	2	2	1	3	2	10
US/ESRO	3	1	0	0	3	0	7

							Total
US/Britain	3	1	1	1	1	0	6
US/Canada	2	1	0	1	1	1	6
US/Italy	2	0	0	1	0	0	3
US/France	1	0	0	1	0	0	2
US/Germany	—	1	0	0	1	0	2
US/Nato	—	—	1	1	0	0	2
USSR/France	—	—	—	1	1	1	3
US/Australia	1	1	1	0	0	0	2
France/Germany	—	—	1	0	0	0	1
Total launches	768	110	114	120	106	109	1327
Total launches still in orbit 1 Apr, 1973	255	34	48	65	59		

Objects in earth orbit: On Aug 31, 1973,
2974 objects (satellites, rocket casings and fragments). Decayed objects 3837.

Note: This log includes manned flights. It is reproduced by permission of the Royal Aircraft Establishment, Farnborough, Hants, England.

Major Unmanned Firsts

Oct 4, 1957	SPUTNIK 1	First artificial satellite
Jan 31, 1958	EXPLORER 1	Discovered Van Allen radiation belts
Mar 17, 1958	VANGUARD 1	Measured earth's shape
Jan 2, 1959	LUNA 1	Earth escape spacecraft
Aug 7, 1959	EXPLORER 6	TV pictures from space
Sep 12, 1959	LUNA 2	Lunar impact
Oct 4, 1959	LUNA 3	Farside lunar picture
Apr 1, 1960	TIROS 1	Weather satellite
May 24, 1960	MIDAS 2	Missile detection satellite
Jun 22, 1960	TRANSIT/SOLRAD	First multiple payload
Aug 10, 1960	DISCOVERER 13	Payload recovered
Feb 12, 1961	SPUTNIK 8	Orbital platform launch
Feb 12, 1961	VENUS 1	Venus fly-by launched by Sputnik 8
Aug 27, 1962	MARINER 2	Data from Venus
Nov 1, 1962	MARS 1	Mars fly-by
Jul 26, 1963	SYNCOM 2	Synchronous satellite
Oct 17, 1963	VELA 1	Satellite to detect nuclear explosions
Jul 28, 1964	RANGER 7	First close-up moon pictures
Nov 28, 1964	MARINER 4	Mars pictures
Apr 3, 1965	SNAPSHOT 1	Nuclear reactor in orbit
Apr 6, 1965	EARLY BIRD (INTELSAT 1)	Commercial TV communications

Date	Spacecraft	Description
Jul 16, 1965	PROTON 1	Cosmic ray measurements
Nov 16, 1965	VENUS 3	Venus impact
Jan 31, 1966	LUNA 9	Soft-landed on moon; pictures from surface
Mar 31, 1966	LUNA 10	Lunar orbiter
Aug 10, 1966	ORBITER 1	Pictures from lunar orbit
Dec 21, 1966	LUNA 13	Tested density of lunar surface
Jan 25, 1967	COSMOS 139	FOBS—Fractional Orbit Bombardment Satellite
Apr 17, 1967	SURVEYOR 3	Dug lunar trench
Jun 12, 1967	VENUS 4	Investigated Venusian atmosphere
Jul 1, 1967	DODGE 1	First full-face colour picture of earth
Sep 8, 1967	SURVEYOR 5	Chemical analysis of lunar soil
Oct 30, 1967	COSMOS 186/188	Automatic docking
Sep 21, 1968	ZOND 5	Circumlunar flight and recovery
Oct 20, 1968	COSMOS 249	Inspection/destructor satellite
Nov 10, 1968	ZOND 6	Skip re-entry after lunar flight
Aug 17, 1970	VENUS 7	Soft-landed on Venus
Sep 12, 1970	LUNA 16	Automatic return of lunar soil
Nov 10, 1970	LUNA 17	Lunar robot
May 19, 1971	MARS 2	Impacted on Mars
May 28, 1971	MARS 3	Soft-landed on Mars
May 30, 1971	MARINER 9	Orbited Mars
Mar 3, 1972	PIONEER 10	Launched to Jupiter
Mar 27, 1972	VENUS 8	Analysed Venusian Soil
Jul 23, 1972	ERTS 1	Earth Resources Satellite

13

Planetary Flights

1 Mars

1 Oct 10, 1960	USSR (U)	Failed to reach earth orbit
2 Oct 14, 1960	USSR (U)	Failed to reach earth orbit
3 Oct 24, 1962	USSR (U)	Exploded in earth orbit
4 Nov 1, 1962	MARS 1	Passed Mars at 124,270 miles (200,000 km) after communications failed
5 Nov 4, 1962	USSR (U)	Failed to leave earth orbit
6 Nov 5, 1964	MARINER 3	No communications when shroud failed to jettison, throwing it off course
7 Nov 28, 1964	MARINER 4	Returned Mars pictures and data, passing at 6116 miles (9844 km) Still active
8 Nov 30, 1964	ZOND 2	Communications failed, but passed Mars at 932 miles (1500 km)
9 Jul 18, 1965	ZOND 3	Engineering test towards orbit of Mars; returned lunar farside pictures en route
10 Feb 24, 1969	MARINER 6	Returned Mars pictures; passed planet on Jul 31, 1969 at 1998 miles (3215 km)

14

11	Mar 27, 1969	MARINER 7	Returned Mars pictures and data; passed planet on Aug 5, 1969 at 2185 miles (3516 km)
12	May 8, 1971	MARINER 8	Failed to achieve orbit
13	May 10, 1971	COSMOS 419	Failed to leave earth orbit
14	May 19, 1971	MARS 2	Gathered data in Martian orbit; lander, carrying USSR coat-of-arms, the hammer-and-sickle emblem, crashed on surface Nov 27, 1971
15	May 28, 1971	MARS 3	Gathered data in Martian orbit; lander survived, touched down, but TV transmissions from surface on Dec 2, 1971 ended after 20 secs
16	May 30, 1971	MARINER 9	Entered Martian orbit on Nov 13, 1971; sent over 7000 TV pictures of the planet and its moons
17	Jul 21, 1973	MARS 4	2 pairs intended for orbital and soft-landing operations
18	Jul 25, 1973	MARS 5	
19	Aug 5, 1973	MARS 6	
20	Aug 9, 1973	MARS 7	

2 Venus

1	Feb 4, 1961	SPUTNIK 7		Earth-orbiting platform failed to launch probe
2	Feb 12, 1961	VENUS 1		Communications failed, but passed Venus at 62,140 miles (100,000 km)
3	Jul 22, 1962	MARINER 1		Destroyed by Range Safety Officer at 100 miles (161 km) altitude
4	Aug 25, 1962	USSR (U)		Failed to leave earth orbit
5	Aug 27, 1962	MARINER 2		Returned data, passing Venus at 21,648 miles (34,830 km)
6	Sep 1, 1962	USSR (U)		Failed to leave earth orbit
7	Sep 12, 1962	USSR (U)		Failed to leave earth orbit
8	Nov 11, 1963	COSMOS 21		Engineering test only; failed to leave earth orbit
9	Mar 27, 1964	COSMOS 27		Failed to leave earth orbit
10	Apr 2, 1964	ZOND 1		Communications failed; passed Venus at 62,140 miles (100,000 km)
11	Nov 12, 1965	VENUS 2		Communications failed; passed Venus at 14,790 miles (23,810 km)
12	Nov 16, 1965	VENUS 3		Communications failed; impacted on Venus
13	Nov 23, 1965	COSMOS 96		Failed to leave earth orbit
14	Jun 12, 1967	VENUS 4		Returned atmospheric data on Oct 18, 1967 during descent and impact

15 Jun 14, 1967	MARINER 5	Returned data during fly-by at 2480 miles (3990 km) on Oct 19, 1967
16 Jun 17, 1967	COSMOS 167	Failed to leave earth orbit
17 Jan 5, 1969	VENUS 5	Returned atmospheric data during descent and impact on May 16, 1969
18 Jan 10, 1969	VENUS 6	Returned atmospheric data during descent and impact on May 17, 1969
19 Aug 17, 1970	VENUS 7	Reached surface and returned data for 23 mins on Dec 15, 1970
20 Aug 22, 1970	COSMOS 359	Insufficient velocity to leave earth orbit
21 Mar 27, 1972	VENUS 8	Reached Venusian surface and returned data for 50 mins on Jul 22, 1972
22 Mar 31, 1972	COSMOS 482	Failed to leave earth orbit
23 Nov 3, 1973	MARINER 10	Venus/Mercury fly-by

3 Jupiter

1 Mar 3, 1972	PIONEER 10	Fastest man-made object; passed Jupiter at distance of 81,000 miles (130,300 km) on Dec 4, 1973
2 Apr 5, 1973	PIONEER 11	Due to pass Jupiter in Dec 1974; may pass Saturn in 1980.

CHINA

History By the end of 1972 China had launched
2 satellites, both from her Shuang-cheng-tzu
launch site in Central China approx. 1000 miles
(1609 km) west of Peking, and near Lop Nor, the
nuclear test base. About the same weight as early
US communications satellites, China's satellites
were believed to have been orbited by a booster
based on Soviet technology—a 2-stage, liquid-
propelled SS–2 Sandal intermediate range ballistic missile (IRBM), capable of placing a 300–
600 lb (136–272 kg) satellite in earth orbit,
depending on launch latitude and inclination.

Development of China's space boosters and
long-range military rockets is largely credited to
Dr Tsien Wei-Ch'ang, who was expelled from the
US in 1949 as a result of the late Senator Joseph
McCarthy's anti-Communist campaigns. Tsien
had taken a Ph.D at Toronto University and then
worked as a research engineer at the Jet Propulsion Laboratory, Pasadena, until ordered to return
to mainland China. He came back to the US as
leader of 7 Chinese scientists in November 1972;
he said China was developing domestic communications satellites, and would launch one 'in
the very near future'.

China 1 L. April 24, 1970 by SS–2 from
Shuang-cheng-tzu. Wt ?381 lb (172 kg). Orbit
274 × 1484 miles (441 × 2386 km). Incl. 68·4°.
China's first satellite made her the fifth space
country. It carried a transmitter which broadcast
'The East is Red', a Chinese song paying tribute
to Chairman Mao, and announcing the times it
passed over various parts of the world. Probably
spheroid, with 3·2 ft (1 m) dia., it transmitted

until June 1970; orbital life, 100 years.

China 2 L. March 3, 1971 by SS–2 from Shuang-cheng-tzu. Wt 485 lb (221 kg). Orbit 169×1134 miles (265×1825 km). Incl. 69·9°. Powered by solar cells, transmitted data for 12 days, when the batteries apparently failed. Orbital life 5 years.

____EUROPEAN SPACE AGENCY

History Formation of a European Space Agency by April 1974, was agreed upon by 11 European countries at a meeting in Brussels on July 31, 1973. The countries were Belgium, Denmark, France, Germany, Italy, Netherlands, Norway, Spain, Sweden, Switzerland and Great Britain. The new Agency would be formed by merging ELDO (European Launcher Development Organization) and ESRO (European Space Research Organization). Its main object was to make better use of the total of £190 million ($456 million) per year being spent by Europe on national and international space programmes. ELDO, with its independent Europa launch capability, was finally abandoned in favour of designing and developing, at a cost of £128 million ($307 million) over an 8-year period, a manned laboratory, which would be an element of NASA's Space Shuttle. Details of this properly belong to the *Observer's Book of Manned Spaceflight*, since Spacelab will enable teams of up to 4 non-astronaut scientists and engineers to work in space for up to a month at a time; but it is interesting to note that since Germany is paying 52% of Spacelab costs, compared with France's 10% and Britain's 6·3%, it seems certain that a West German will fill the one place provided on the first flight, around 1980, for Europe's first astronaut. ESA's unmanned activities include development of the French L3S launcher at a cost of £190 million ($456 million), over the same 8-year period, and of the British-backed MAROTS (Maritime Orbital Test Satellites), at a cost of £31 million ($74·5 million), over a 6-year period.

ELDO

ELDO The 7-nation European Launcher Development Organization was set up on February 29, 1964. The initiative came from Britain, anxious to find a use for Blue Streak, originally developed as a strategic missile, as the first stage of an independent European launcher system. Member countries were Belgium, France, West Germany, Italy, Netherlands, and Great Britain, plus Australia; her Government was equally anxious to keep the Woomera Test Range, north of Adelaide, in business. With France providing the 2nd stage, Coralie; Germany the 3rd stage, Astris; Italy the satellite test vehicles, Belgium the ground guidance station; and Netherlands, telemetry links and other equipment, Europa 1 was built. By June 12, 1970 10 test firings had been carried out at Woomera; despite a series of earlier failures, there was confidence that the last, F-9, would place a test satellite in orbit. But loss of thrust on the 3rd stage, and failure to jettison the nose cone, prevented this. Despite growing doubts about the project, and Britain's decision to withdraw, announced in 1970, ELDO pressed ahead with development of the more advanced Europa 2 vehicle, which had the addition of a French-built, 4th stage perigee motor, able to place a satellite of up to 440 lb (200 kg) in stationary orbit. The objective was to launch 2 Franco-German Symphonie telecommunications satellites in 1973 and 1974. Unfortunately the first test-firing of Europa 2, from Kourou on November 5, 1971, with the world's space correspondents observing, ended in disaster. After $2\frac{1}{2}$ mins, it deviated from its flight trajectory, became overstressed, and exploded. Amid much bitter argument, plans continued for the next test launch, F-12, and the vehicle was aboard a French ship en route to Kourou, when on April 27, 1973 the project was finally abandoned in favour of creating a European Space Agency, and building Spacelab for America's Space Shuttle system. Up to that time, Europe had spent about £310 million ($745 million) on ELDO.

ESRO The 10-nation European Space Research Organization was set up in 1964 to promote European collaboration in space research and technology. Member states are Belgium, Denmark, France, West Germany, Italy, Netherlands, Spain, Sweden, Switzerland, and Great Britain. With headquarters in Paris, and pay-roll of 1100, its 1973 income was £53 million ($127·5 million). By 1972 American launchers had successfully placed 7 ESRO satellites in orbit, with 4 more due for launch in 1975–79.

FRANCE

History Having consistently spent more on space than any country after Russia and America, France became the third nation to achieve a national orbital capability, when the 18-ton, 3-stage Diamant launcher placed the A-1 satellite in orbit on November 26, 1965, 8 years after Sputnik 1. By the end of 1972 France had orbited 7 satellites (the fifth being a West German payload). In addition, she had arranged for 2 of her own research satellites to be orbited by America (FR-1 on

December 6, 1965; and EOLE on August 16, 1971) and 2 more, to study the upper atmosphere and environment, by Russia (Aureole, on December 27, 1971, and SRET launched with Molniya 1 on April 4, 1972). France has been remarkably successful in collaborating with both the big space powers at the same time. In addition to the launch agreements, her scientists have contributed important experiments to major space projects of both powers; examples include the laser mirrors on Russia's Lunokhod vehicles, experiments on the 1973 Mars probes, and the ultra-violet telescope to measure star brightness included in the US Skylab space station. The final collapse of ELDO in 1973, however, in which France was the driving force, coincided with a run of misfortunes in her national programme.

Diamant Launcher The 'A' version has an overall length of 61 ft 7 in (18·75 m), 1st stage dia. 4 ft 7 in (1·40 m), and wt, with payload of 39,616 lb (17,970 kg). It can place 175 lb (80 kg) in a 185-mile (300-km) orbit. The 1st stage, Emeraude, was first launched, without complete success, on June 17, 1964, but 21 subsequent firings were all successful; its Vexin liquid-propellant engine provides 61,730 lb (28,000 kg) thrust for 88 secs through a single gimballed chamber. 2nd stage, Topaze, has a polyurethane-type solid propellant motor in a steel casing, providing 31,967 lb (14,500 kg) thrust for 39 secs through 4 gimballed nozzles. 3rd stage, with similar solid propellant motor in glass-fibre wound casing, provides a variable thrust of 5512–11,685 lb (2500–5300 kg) thrust for 44·5 secs through a single fixed nozzle. This rocket achieved the first 3 orbital launchings outside America and the Soviet Union.

Diamant B, 77 ft 3 in. (23·54 m) long, and launch

wt of 54,235 lb (24,600 kg), has an L17 1st stage, with liquid bi-propellant motor, developing 77,160 lb (35,000 kg) thrust at lift-off for 112 secs. 2nd and 3rd stages are similar to Diamant A. First launching of Diamant B was at Kourou, on March 10, 1970, when it placed France's first foreign payload in orbit. Future versions are planned with a 4th stage, giving it the ability to place 265 lb (120 kg) into a circular 1000-mile (1600-km) orbit.

Guiana Space Centre France began her national space programme with launches from Hammaguir, in the Sahara Desert, but in 1964 decided to create a civil launch centre at Kourou, in French Guiana. Though hot and wet, its position 2° north of the Equator makes it ideal for launching synchronous satellites, and to the north and east rockets can travel 1860 miles (3000 km) without passing over land. The first firing from Guiana was a sounding rocket on April 9, 1968, with the first satellite being launched in March, 1970. With the decision in 1966 to transfer ELDO (European Launcher Development Organization) activities from Woomera in Australia to the more favourable Guiana site, it was developed into a space centre comparable with America's Cape Kennedy, and in some respects more modern. The final collapse of ELDO in 1973, however, left the Guiana Centre with only one or two Diamant B launches per year, and the French having to bear the whole of the £2·4 million ($5·7 million) per year upkeep costs, instead of being able to rely on a 40% contribution from ELDO.

National Launches

A-1 (Asterix) L. Nov 26, 1965 by Diamant from Hammaguir. Wt 92 lb (41·7 kg). Orbit 328 × 1099 mi (528 × 1768 km). Incl. 34°. First French satellite. Transmitted for 2 days. Orbital life 200 yrs.

D-1A (Diapason) L. Feb 17, 1966 by Diamant from Hammaguir. Wt 44 lb (20 kg). Orbit 313 × 1711 mi (504 × 2753 km). Incl. 34°. 3rd French satellite and 2nd national launch. Carried out geodetic data research. Orbital life 200 yrs. **D-1C (Diademe 1)** L. Feb 8, 1967 by Diamant from Hammaguir. Wt 50 lb (22·6 kg). Orbit 360 × 833 mi (580 × 1340 km). Incl. 40°. Geodetic satellite, operational in spite of low apogee. Orbital life 100 yrs. **D-1D (Diademe 2)** L. Feb 15, 1967 by Diamant from Hammaguir. Wt 50 lb (22·6 kg). Orbit 368 × 1172 mi (592 × 1886 km). Incl. 39°. Provided laser and doppler data for 3 months; orbital life 200 yrs. **Wika/Mika** L. Mar 10, 1970 by Diamant B from Kourou. Wt Wika 110 lb (50 kg). Mika 143 lb (64·8 kg). Orbits 191 × 1002 & 1029 mi (307 × 1700 & 1746 km). Incl. 5°. The W. German Wika satellite was the first foreign payload to be orbited by France; it carried out 30 days of geocorona and upper atmosphere investigation, and had a 5-yr orbital life. Mika's task was to monitor the performance of the Diamant B launcher, which had 34,000 lb (15,420 kg) thrust, liquid 1st stage, solid propellant 2nd stage and 3rd stage apogee motor to spin payload into orbit. **Peole** L. Dec 12, 1970 by Diamant B from Kourou. Wt 154 lb (69·7 kg). Orbit 360 × 464 mi (580 × 747 km). Incl. 15°. Test flight to qualify the EOLE meteorology satellite, successfully launched by Scout B from Wallops, Va. on Aug 16, 1971. This project, being conducted in collaboration with NASA, involves using a satellite to track hundreds of 13-ft (4-m) pressurized meteorological balloons, carrying capsules on cables and measuring the temperatures surrounding them, etc., as they drift at a constant height. **D-2A (Tournesol 1)** L. Apr 15, 1971 by Diamant B from Kourou. Wt 200 lb (90 kg). Orbit 283 × 437 mi (456 × 703 km). Incl. 46°. Carried 5 experiments to study solar radiation and ultra-violet range. Orbital life 6 yrs. **D-2A** L. Dec 5, 1971 by Diamant B from Kourou. Wt 214 lb (97 kg). Intended for a higher orbit than the previous satellite, to continue solar radiation studies, but 2nd stage of launcher failed. **D-2A** L. May 21, 1973 by Diamant B from Kourou. Twin satellites, Castor and Pollux, fell into the sea; apparently the launcher failed to produce sufficient thrust.

GREAT BRITAIN

History Britain's first nationally launched satellite was also her last. Political hesitation and

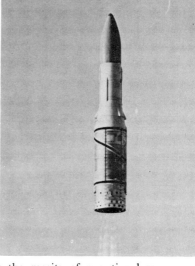

Britain's Black Arrow being tested at Woomera, Australia, for launch of Prospero

vacillation over the merits of a national space programme go back (like the US and Soviet programmes) to ballistic missile development. In Britain this was the Hawker Siddeley Blue Streak, finally cancelled as a missile in 1960, though continued until 1973 as a 1st stage launcher for the European ELDO programme. A more modest attempt at a national launch vehicle was announced in September 1964. It was decided to develop a small 3-stage vehicle from the successful Westland Black Knight research rocket. Construction and supply of 3 launchers, to be called Black Arrow, with the first 2 stages directly evolved from Black Knight, was ordered in March 1967. The purpose was to develop and test in space new components for communications satellites; and to develop a tool for space research. The successful accomplishment of these objectives in October 1971 was inevitably soured by the fact that in July the Government had announced that the programme would be cancelled after the final launch. But with this one satellite Britain became the sixth nation to achieve orbital capability—after Russia, America, France, Japan and China.

In addition to the national space programme, a total of 6 satellites had been launched for Britain by NASA by the end of 1973. Ariels 1–4 (also known as UK 1–4), a series of small research satellites studying electron density, terrestrial radio noise etc., in a joint US/UK programme, were launched on April 26, 1962, March 27, 1964, May 5, 1967 and December 11, 1971. For details of 2 military satellites, Skynet 1 and 2, see US Military Satellites (Communications).

BLACK ARROW/PROSPERO

Test flights and the final launch result were as follows:

X-1 L. June 27, 1969 from Woomera. Veered off course, and had to be destroyed 50 secs after lift-off. Intended to provide data on 1st and 2nd stage performance, with separation of dummy third stage.

R-1 L. March 4, 1970. Completely successful test of 1st and 2nd-stage engines, and of 3rd-stage systems (propulsion test not included), and spin-up of dummy payload. 2nd stage and payload impacted as planned, 15 mins after launch, in Indian Ocean, 1900 miles (3050 km) NW of Woomera.

R-2 L. September 1970. Intended to prove 3rd-stage propulsion system and inject into orbit X–2 development payload. First objective achieved but orbit not achieved, due to failure to maintain 2nd-stage (HTP) tank pressurization.

R-3 Prospero L. October 28, 1971, by Black Arrow from Woomera, Australia. Wt 145 lb (66 kg). Orbit 340 × 983 miles (547 × 1582 km). Incl. 82°. Design aim of 1-year life more than achieved; 812 passes were monitored from total of 4960 orbits in first year, and on-board tape-recorder replayed 697 times. It failed on May 24, 1973 after 730 replays; but after 2 years in orbit, real-time data was still being received at the Lasham, Hants, telemetry station, at the rate of 1 pass per week. Orbital life of satellite, 150 years; rocket, 100 years.

Launcher Description Overall ht 42 ft 4 in. (12·9 m). Dia. 1st stage, 6 ft 6¾ in. (2 m). 1st stage: liquid-fuelled (HTP and kerosene) engine is Rolls Royce Bristol Gamma 8 with cluster of 8 thrust chambers, swivelled in 4 pairs for control in pitch, yaw and roll. Thrust 50,000 lb (22,680 kg). 2nd stage: 2 RR Bristol Gamma 2 thrust chambers, also liquid fuelled, and fully gimballed for

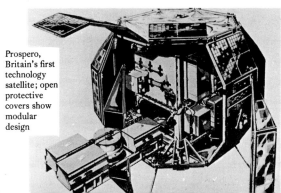

Prospero, Britain's first technology satellite; open protective covers show modular design

flight control. Thrust 15,300 lb (6940 kg). 3rd stage: solid propellant motor developed from 2nd stage of Black Knight. This and payload are mounted within protective fairings which are jettisoned early in 2nd-stage firing. Guidance is fixed-programme, with gyro system maintaining course; by end of 2nd-stage thrust, the vehicle can coast up to orbital injection point, achieved by firing 3rd stage.

Prospero Description Wt 145 lb (66 kg). Ht 28 in. (71 cm). Dia. 45 in. (114 cm). Pumpkin shaped. Intended to prove new telemetry and power-supply system for satellites with electrical load of less than 30W; to test thermal surfacing, lightweight solar cells, etc. These are carried on 8 large and 8 small segments attached to main box section. 4 telemetry aerials mounted on base at 45° to spin axis. Internal equipment includes telemetry and data handling, micrometeoroid detector, electronics, etc.

JAPAN

History Japan was the fourth country, after the Soviet Union, United States and France, to achieve satellite capability. When Osumi, the first test satellite, was orbited in 1971 Japan's space budget over a period of 16 years totalled only £78.7 million ($189 million). That initial success, however, was 2½ years later than had been hoped, since the first unsuccessful attempt was made in September 1966. The first true satellite followed in 1971, 3 years later than planned. These delays, which robbed Japan's scientists of their plans to become the third space nation, may have been partly because Japan had 2 rival space programmes

—with Tokyo University working on solid-fuel rockets, and the Government's Science and Technology Agency on liquid-fuelled rockets. This was resolved with the setting up on October 1, 1969, of NASDA (National Space Development Agency) to promote development of satellites, launchers, tracking systems, etc. NASDA is now preparing to launch a number of scientific satellites in 1975–77, culminating with an experimental communications satellite. A new 3-stage rocket, designated 'N', is being developed, with lift-off thrust of 330,600 lb (150,000 kg), capable of putting a 220 lb (100 kg) satellite into stationary orbit. With a soundly based and financed long-term space programme, Japan is now regarded as having the best chance of becoming the world's third nation to develop the capability of putting men into space.

Osumi L. February 11, 1970 by Lambda 4S from Kagoshima Space Centre. Orbit 210 × 3200 miles (338 × 5150 km). Incl. 31°. Named after the district of Kyushu in which the space centre is located, Osumi was in fact the 4th stage of the rocket, containing a 26-lb (12-kg) instrument package, as test satellite. Its telemetry, pilot and beacons transmitter systems worked normally for 20 hrs. It was the 5th flight of the solid-propellant Lambda 4S, consisting of 1st stage core with 81,500 lb (36,970 kg) thrust, augmented by 2 29,000 lb (13,150 kg) strap-on boosters; 2nd stage providing 26,000 lb (11,800 kg) thrust; 3rd stage with 14,500 lb (6580 kg) thrust; and 4th stage with 1800 lb (816 kg) thrust. Ht 54 ft (16 m); wt 20,700 lb (9390 kg). First 3 stages were unguided; 4th stage was provided with attitude control only.

Tansei L. February 16, 1971 by MU–4S from Kagoshima. Wt 136 lb (62 kg). Orbit 615 × 690 miles (990 × 1110 km). Incl. 30°. Named Tansei ('Light Blue'), after the colours of Tokyo University, the spherical satellite was a test payload, carrying thermometer, accelerometer and transmitter, which operated for its full lifetime of 1 week. Orbital life, 1000 years. Main purpose of the launch was to demonstrate the capability of the MU–4S 4-stage solid fuel rocket; this had 1st stage thrust of 187,000 lb (84,820 kg) and was 77 ft (23·4 m) high.

Shinsei L. September 28, 1971 by MU–4S from Kagoshima. Wt 143 lb (65 kg). Orbit 543 × 1162 miles (874 × 1870 km). Incl. 32°. Japan's first scientific satellite, named Shinsei, or 'New Star' to mark the occasion. 4 ft (1·2 m) high, with solar arrays mounted on an octahedral structure, it measured the ionosphere, solar and cosmic radiation. Orbital life, 5000 years.

First Delta-N launcher, showing 6 solid-fuel strap-on rockets for additional lift-off thrust

LAUNCHERS

The following is a very brief summary of US Rockets used to launch unmanned satellites:

Agena Versatile upper stage, used on top of Thor, Atlas, Thrust Augmented Thor (TAT), Long-Tank Thor, Thorad, and Titan 3B. Typical length, 23 ft 3 in. (7·09 m). Thrust 16,000 lb (7257 kg).

Atlas/Agena General-purpose launcher developed in various versions from original ICBM, used by both USAF and NASA. Lift-off thrust about 350,000 lb (158,760 kg). Overall ht 104 ft (31·7 m).

Atlas/Centaur High-energy, 2-stage rocket, able to lift medium-wt spacecraft of 9900 lb (4500 kg) into 345-mile (555-km) orbits, or 4000 lb (1810 kg), into synchronous transfer orbits. Also used for planetary missions. Atlas provides 431,000 lb (195,500 kg) lift-off thrust; overall ht 117·4 ft (35·8 m); max. dia. 10 ft (3·04 m). By end of 1972 21 Atlas/Centaurs had been launched with 86% success rate.

Centaur Used as upper stage on Atlas and Titan launchers. First US space vehicle to use liquid hydrogen as a propellant; capable of putting 4·5 ton (4570 kg) payloads into orbit, and sending large payloads to planets. Thrust 30,000 lb (13,610 kg). Ht 30 ft (9·14 m).

Delta Versatile launcher based on Thor 1st stage, used for wide variety of medium-sized satellites and small space probes. Can be used as 2 or 3-stages, augmented with 3, 6 or 9 solid-fuel strap-on 1st-stage motors. Thor 1st stage provides 478,000 lb (216,800 kg) thrust, including 6 strap-ons. Overall ht 115 ft (35 m); max dia. 8 ft (2·4 m). By end of 1972 93 Deltas had been launched with 91% success rate.

Juno   *See* Explorer entry, page 46.

Jupiter   *See* Explorer entry, page 46.

LTTAD   Long Tank Thrust Augmented Delta. Modified Thor 1st stage, with 17,200 lb (7800 kg) thrust, augmented by 3, 6 or 9 strap-on solid propellant rockets giving combined thrust of 335,500 lb (152,200 kg). Ht 105 ft (31·9 m) approx. Able to launch 350 lb (158 kg) to stationary orbits.

Scout   America's smallest and only solid-fuel launcher, used for large variety of small scientific payloads, and high-speed re-entry experiments. 4 stages; with lift-off thrust of 107,100 lb (48,580 kg). Ht 73 ft (22 m), max dia. 3·67 ft (1·12 m), can place 420 lb (190 kg) in 300-mile (480-km) orbit due east from Wallops Island, Virginia. By end of 1972, 60 Scouts had been launched with 95% success rate.

TAT   Thrust Augmented Thor.

Thor   Originally America's first IRBM, now provides, in various forms, her most frequently used launcher. Fuelled by liquid oxygen and kerosene. Starting in 1957 with a 1st-stage thrust of 150,000 lb (68,000 kg), the latest Long Tank Thor provides 330,000 lb (149,700 kg). Ht 70 ft 6 in. (21·5 m). Constant dia. 8 ft (2·4 m). Provides 1st stage of Delta series; by end of 1972 about 450 Thors had been launched.

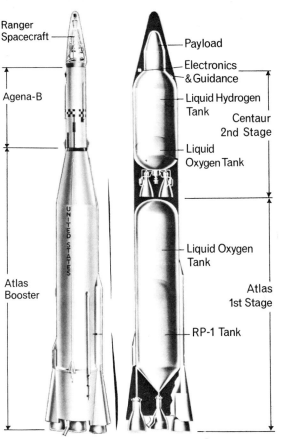

Ranger Spacecraft

Agena-B

Atlas Booster

UNITED STATES

Payload

Electronics & Guidance

Liquid Hydrogen Tank

Centaur 2nd Stage

Liquid Oxygen Tank

Liquid Oxygen Tank

Atlas 1st Stage

RP-1 Tank

20 out of 26 Atlas-Agenas have been successful

24 out of 30 Atlas-Centaur firings have succeeded

Thorad/Agena D Uprated version of TAD (Thrust Augmented Delta) with Agena 2nd stage. 1st stage, with 3 strap-on solid motors, gives lift-off thrust of 326,000 lb (147,870 kg). Ht 109 ft (33 m).

Titan 3C USAF vehicle, mainly for military launches. 4-stages. 2 strap-on solid fuel rockets provide 2,300,000 lb (1,043,300 kg) lift-off thrust, followed by 485,000 lb (220,000 kg) thrust from main core booster. Can place 23,000 lb (10,570 kg) into 345-mile (555-km) orbits, or 8000 lb (3628 kg) into synchronous orbits; or send 2650 lb (1202 kg) to Mars or Venus. Ht of 4-stage Titan 3C 127 ft (38·7 m). Max dia. 10 ft (3 m). With Centaur 4th stage, overall ht 160 ft (48·7 m); can send 7600 lb (3450 kg) to planets (*see* page 115).

Vanguard *See* Vanguard entry, pages 112–114.

DISCOVERER

History An early programme of major importance. Its tests on orbital manoeuvring and re-entry techniques not only played a large part in enabling the first manned flights to be made in Project Mercury, but developed the 'spy satellite' systems now in regular use. Between 1959 and 1962, after which this type of work was classified, there were 38 launches; all were made by the US Air Force from the Western Test Range at Vandenberg. Project objectives were: 'military space research; development of capsule recovery techniques; and biological research.' The series began 13 months after the successful launch of Explorer 1, America's first satellite, but for 18 months after that, as the list of the early flights

Scout 5 ready for
launch at Wallops,
Va., in June 1961

Thor rocket, with Echo balloon
on top, awaiting launch on
Cape Canaveral's Pad 17

on the following pages illustrates, it was dogged by failures. The plan was for a modified Thor IRBM to boost a 2nd-stage Agena A rocket to near-orbit altitude; after separation, the Agena's 15,000-lb (6804-kg) engine was fired to place the whole stage in orbit. There its gas-jet orientation system turned the vehicle through 180°, so that the ballistic nose-cone containing a 300-lb (136-kg) ejection capsule, was pointing rearwards and tilted down. It thus met the requirement that, for a satellite to drop a bomb or return an object to earth, the object must separate and use retro-rockets fired in the opposite direction to the satellite's path, at a downward angle to ensure that it re-enters the atmosphere. At first ejections were often at the wrong angle, and sometimes upwards. This resulted in a false alarm in February 1960, when the US Navy's early detection system found an unknown satellite, believed to be Russian, in a near-Polar orbit. It was ultimately established that it was an early Discoverer cap-sule which had been ejected upwards instead of down. Ejection was achieved by explosive bolts, on either orbit 17 or 33. At 50,000 ft (15,240 m) a parachute pulled the recovery package clear of the heat-shield, which fell into the sea—in these early tests usually in the Pacific near Hawaii. Radio and light beacons, and even radar chaff, were used to assist location as the equipment package descended into the target area. At first C119 transport air-craft, towing trapeze-like frameworks at a height of 8000 ft (2438 m), patrolled the area in some numbers in vain efforts to develop the technique of 'snatching' the parachute harness with the package attached, before it reached the sea. A sea recovery was finally achieved on Explorer 13, and the first mid-air capture on the following mission. Beginning with Discoverer 30 in September 1961, C130s were used in this role. Amid much public concern that America's military capacity in space was probably far behind Russia's, the programme was pursued with almost frantic haste. There were 12 launches in 1961, with the last 2 officially designated Discoverers being launched in 1962. Looking back, and noting the large percentage of failures, it is easy to forget that space computer techniques, as well as hardware such as re-entry nose-cones, were still being developed. But in the 21 launches between the first mid-air success and Discoverer 38, there were only 7 more successful mid-air recoveries; 3 sea recoveries could be counted partial successes; 11 launches were admitted failures. The other 3 could probably be counted as experimental. The high proportion of failures inevitably attracted much public comment; Russia meanwhile was able to pursue her parallel programme in complete secrecy, merely announcing the Cosmos series numbers of her launches. Inevitably, from November 22, 1961 the US Department of Defense decided to classify all military space-flights. Since then it has been possible to follow such activities only indirectly by studying the Tables of Earth Satellites, compiled by Britain's Royal Aircraft Establishment. These show 27 capsule ejections in orbit between 1963–71; undesignated launchings are listed under the name of their rocket, Atlas–Agenas B and D until 1967, when more powerful Thor–Agenas and Titan–Agenas began to come into use. These launches have long since passed from the experimental to the operational stage; further details can be found under the heading MILITARY SATELLITES of the regular passes now being made over Russia, China, Middle East and other countries, whether or not they happen to be US allies, by satellites able to return data either directly or by means of ejected film packs. Below is a detailed list of the early Discoverers and, for comparison, of the last of them:

Discoverer 1 L. Feb 28, 1959 by Thor–Agena A from Vandenberg. Wt 1300 lb (590 kg), including 245 lb (111 kg) of instruments. Orbit 99 × 605 mi (159 × 974 km). Incl. 90°. First satellite in polar orbit. Objective was to test propulsion, guidance, staging and communications, but accurate tracking was prevented by tumbling. Re-entered after 5 days. **Discoverer 2** L. Apr 13, 1959 by Thor–Agena A from Vandenberg. Wt 1610 lb (730 kg). Orbit 142 × 220 mi (229 × 354 km). Incl. 90°. Objective was to eject and recover a 195-lb (88-kg) hemispherical capsule, inside which temperature and oxygen would be sufficient to maintain life. Capsule successfully ejected on orbit 17, but was lost in Arctic. Satellite re-entered after 13 days. **Discoverer 3** L. Jun 3, 1959 by Thor–Agena A from Vandenberg. Similar to Discoverer 2, but carrying 4 black mice in recoverable capsule. Failed to orbit after 2nd-stage failure. **Discoverer 4** L. Jun 25, 1959. Similar to Discoverer 2. Failed to orbit, again because of 2nd-stage failure. **Discoverer 5** L. Aug 13, 1959. Similar to Discoverer 2. Orbit 136 × 450 mi (219 × 724 km). Incl. 80°. Capsule ejected, but not recovered. Satellite decayed Feb 11, 1961. **Discoverer 6** L. Aug 19, 1959. Orbit 131 × 528 mi (210 × 850 km). Incl. 84°. Capsule again ejected on orbit 17, but recovery failed. **Discoverer 7** L. Nov 7, 1959. Orbit 99 × 519 mi (159 × 835 km). Incl. 81°. Due to poor stabilization, capsule was not ejected. Decayed after 19 days. **Discoverer 8** L. Nov 20, 1959. Orbit 120 × 1032 mi (193 × 1660 km). Incl. 80°. Capsule ejected on orbit 15, but overshot recovery area. Decayed Mar 8, 1960. **Discoverer 9** L. Feb 4, 1960. Failed to orbit; premature 1st-stage cut-off. **Discoverer 10** L. Feb 19, 1960. Failed to orbit; destroyed by Range Safety Officer when it went off course. **Discoverer 11** L. Apr 15, 1960. Orbit 103 × 375 mi (165 × 603 km). Incl. 80°. Capsule was ejected on orbit 17, but recovery failed. Decayed after 11 days. **Discoverer 12** L. Jun 29, 1960. Failed to orbit following 2nd-stage attitude instability. **Discoverer 13** L. Aug 10, 1960. Orbit 157 × 431 mi (253 × 694 km). Incl. 82°. Success at last; capsule ejected on orbit 17, and recovered from sea. Satellite decayed Dec 14, 1960. **Discoverer 14** L. Aug 18, 1960. Orbit 113 × 502 mi (182 × 808 km). Incl. 79°. After ejection on orbit 17, capsule was captured for the first time in mid-air by patrolling aircraft. Decayed after 28 days. **Discoverer 38** L. Feb 27, 1962 by Thor–Agena B from Vandenberg. Wt 2100 lb (952 kg). Orbit 208 × 308 mi (334 × 495 km). Incl. 82°. Last in the series to be officially announced; capsule was successfully ejected and recovered in mid-air after 65 orbits. Satellite decayed Mar 21, 1962.

ERTS—EARTH RESOURCES

History Despite the fact that it was only fully operational for 8 months, instead of the planned minimum of 1 year, ERTS 1 provided the US with the most remarkable success yet achieved with this type of satellite. Earth Resources Technology Satellites are improved and enlarged versions of the Nimbus weather satellites, which they resemble in appearance. The original plan was to launch 2 in successive years (1972/3) as an experiment in systematically surveying the earth's surface to study the health of its crops, and the potential use and development of its land and oceans. The dramatic flow of vivid, revealing photographs (actually 'false-colour images') sent back from the moment it became operational, were regarded as sensational by 300 investigators in the 50 countries participating in the experiment; their quality, clarity and visibility led to the information being put to immediate practical use. Examples of the pictures can be found in the colour section of this book.

Before launch, NASA decreed that all ERTS data, including 9000 pictures per week, should be unclassified and made available to the public; the photographs can be bought for about £0.50 ($1.25). The immediate success of the project led to the processing and distribution systems being swamped with requests. Just as Nimbus experience was used for the development of

ERTS, knowledge gained from ERTS was used, less than a year later, by the rotating crews of astronauts operating even more advanced equipment in Skylab. Total ERTS costs are £72.5 million ($174 million) including £11.6 million ($28 million) for data-handling computers and processing facilities, and £14 million ($34 million) for investigations.

Spacecraft Description Butterfly-shaped. Ht 10 ft (3 m). Width 11 ft (3·4 m) with solar paddles deployed. Sensors and electronics are housed inside a 5-ft (1·5-m) dia. sensory ring below the paddles. Sensors as follows. MSS (Multispectral Scanner Subsystem): this collects data by continually scanning the ground directly below in 4 spectral bands, 2 in the visible spectrum and 2 in the near infra-red. RBV (Return Beam Vidicon Subsystem): this views the same 115-mile (185-km) swathe as the MSS. 3 cameras of 4125 lines are reshuttered simultaneously every 25 secs to produce 115×115-mile (185×185-km) overlapping images of the ground along the direction of the satellite motion. WBVTR (Wide Band Video Tape Recorder): because ERTS is not within range of a ground station for much of its time, 2 tape-recorder systems can store images, and later transmit them, simultaneously. Data Collection System: this collects measurements as it passes over 'remote platforms' installed on icebergs, etc. to send up readings of soil, water and other conditions for later retransmission and comparison with data collected from the other sensors.

ERTS-1 L. July 23, 1972 by Thor–Delta from Point Arguello. Wt 1965 lb (891 kg). Orbit 560×572 miles (901×920 km). Incl. 99°. The near-polar orbit enables the satellite to circle the earth 14

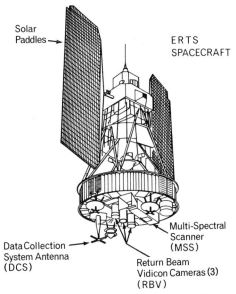

Solar
Paddles

E R T S
SPACECRAFT

Multi-Spectral
Scanner
(MSS)

Data Collection
System Antenna
(DCS)

Return Beam
Vidicon Cameras (3)
(RBV)

times a day, and to view any point on earth except
for small areas around the Poles. Because the orbit
is also sun-synchronous, every 18 days the ERTS
cameras can view the same spot at the same time of
day; each pass enables the cameras to view a
swathe 115 miles (185 km) wide with some overlap
on each pass; it is back in its original position
after 252 passes. By March 30, 1973, when faults
occurred in the video tape-recorder, the N.
American continent had been photographed 10
times; the total of over 34,000 images taken since
launch included all the world's major land masses
at least once. The tape-recorder fault meant that
images could no longer be stored for transmission;

but when the satellite was over one of NASA's 3 ground receiving stations in California, Alaska and Maryland, it was still possible to transmit 'live' pictures of the entire N. American continent. Orbital life 100 years.

ERTS-2 Launch, due in 1973, postponed following budget economies, possibly until 1976.

Remote Sensing The basis of ERTS is that all objects, living or inanimate, transmit or reflect visible and invisible light, and thus have their own 'signature' or individual 'fingerprint'. All energy coming to earth from the' sun is either reflected, transmitted, or absorbed in objects on earth, each in its own way. 'Remote sensing'—measuring objects from a distance—goes back to aerial photography in the 1930s. It began to develop more rapidly with military reconnaissance in World War II, followed by the use of sounding rockets and satellites after that. A camera is a remote sensor in that it records the shape and colour of an object by its reflected light without touching the object. The human eye is the simplest example of remote sensing; but most of the information reflected or radiated by the earth, cannot be detected by the human eye. Near infra-red light lies just beyond our vision. Healthy green vegetation is even brighter in the infra-red than in the visible, and this information is particularly valuable to farmers, because it can provide early warning if their crops are sick. Just as the eye can cover only a minute portion of the total electro-magnetic spectrum, so no single instrument is capable of sensing and measuring the radiations from objects with different physical and chemical properties. One of the objects of ERTS is to determine what sort of sensors, and what combination of them, will yield the most useful information.

Picture Reading and Results The energy reflections described above are converted by the ERTS scanners into electrical signals in 4 selected bands. (In this case, Bands 1 and 2 are in the visible wavelengths of 0·5–0·6 and 0·6–0·7 micrometres; Bands 3 and 4, which are not visible to the human eye, are in the near infra-red portion of the spectrum, with wavelengths of 0·7–0·8 and 0·8–1·1 micrometres). These reflections are processed into 'digital bits', transmitted to a receiving station, and then reprocessed into either black-and-white reproductions of what was seen on earth, or as 'false colour images', by projecting the

This ERTS picture of SE England revealed for the first time a geological 'fault'

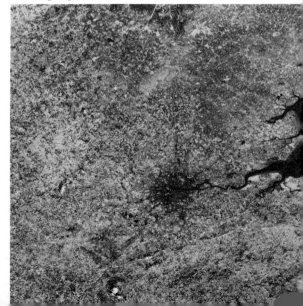

data of 3 of the 4 bands through blue, red and green filters. The colours assigned are in the same order as the primary colours of the visible spectrum, but result in an 'exchanged' colour: what we see as green in Band 1 is shown as blue; what we see as red in Band 2 appears as green; and Band 3, which normally we cannot see, appears as red. Band 4 may be used instead of Band 3, and also appears red.

The result of this is that clear water will appear black in Band 3 and 4, because water almost totally absorbs radiant energy—in other words sends back hardly any reflections; water carrying silt, or otherwise polluted, will appear blue. Trees and plants appear bright red because of the very high reflectivity of chlorophyll-bearing leaves in the near infra-red; vegetation brightness depends on such things as the size of leaves, big leaves showing up as brighter than small leaves, with the effect that hardwood trees will register as brighter than pine trees. The big leaves of tobacco plants will be brighter than wheat. And, the vital factor in the whole exercise, crop brightness depends on plant health; thus, healthy crops, shown in the infra-red Bands 3 and 4, will be much brighter than diseased vegetation. But crop disease would be difficult or impossible to detect on a single photograph; abnormal changes, suggesting disease, would show up when successive pictures were compared.

The amount of knowledge which can be extracted from the data and pictures will increase as the scientists learn the technique of reading them. But at an early state, observers were able to detect geologic faults and water-bearing rock areas in Nebraska, Illinois, and New York State, which had been unknown before; areas of clear and polluted waters in Chesapeake Bay were readily discernible.

From Brazil came reports that the ERTS pictures had revealed that villages and towns were sometimes wrongly located on their maps by tens of kilometres, and that lagoons shown as 20 km long were in reality over 100 km long. Ghana reported that locust-control was being attempted because pictures had shown vegetation at the edge of deserts which attracted locusts for breeding. Iran reported that it had become apparent that lowering of the Caspian Sea by evaporation had apparently changed the shape of the Bandar Shah peninsula. Pictures of Britain are acclaimed as 'remarkable', and disclose, among other things, that a linear feature or 'fault' starts near Harwich on the E. coast and runs right through London to Land's End.

EXPLORER

History   Explorer 1 became the United States' first satellite, and the world's third. The Explorer series, which still continues, is somewhat comparable to Russia's Cosmos in that it embraces a varied series of experimental, research and scientific satellites.

The series began in a flurry of haste, following the successful Soviet launch of Sputnik 1 on October 4, 1957. While the Russians had fulfilled their undertaking, made 2 years earlier, to launch a satellite for meteorological purposes as part of the International Geophysical Year of 1957/58, America's own efforts, concentrated on the Vanguard project (*see* page 112), had been a dismal failure. It was at this point that Washington turned at last to Dr Wernher von Braun's group, at the Army Ballistic Missile Agency at Huntsville, whose satellite proposals had been repeatedly turned down. They had evolved a rocket called Juno 1—a 4-stage development of Jupiter C, which itself had been developed from Redstone, the 70-ft (21·3 m) rocket which von Braun had evolved for America from his wartime German V2. In a plan submitted in April, 1957, ABMA had recommended a programme to launch 6 17-lb (7·7-kg) satellites, the first of which would orbit

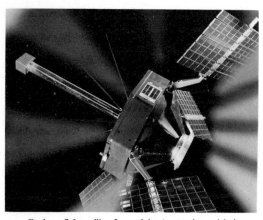

Explorer S-3 satellite, for studying 'energetic particles'

in September 1957. Although of course, it could not have been foreseen, this would have given America a 1-month lead over Russia in what was to become the major event of the 20th century— the 'Space Race'. 3 weeks after Sputnik 1, the ABMA group was given authority to go ahead with plans to launch 2 satellites, with a target date of January 30, 1958 for the first. Explorer 1 was successfully launched 1 day behind this schedule. To its technical success can be added its major contribution to the International Geophysical Year—confirmation of the existence of radiation belts around the earth, forecast and named after Dr James Van Allen. While the early Explorers were tiny compared with Russia's Sputniks, their miniaturized instruments gathered data of extreme scientific value; it was a trend which ultimately put America far ahead in space techniques, and ensured, in spite of Russia's apparent lead in the early years, that the first men on the moon were Americans. Details of some of the Explorer flights are listed below:

Explorer 1 L. Jan 31, 1958 by Jupiter C from Cape Canaveral. Wt (including integral last-stage motor) 30·8 lb (14 kg). Orbit 224 × 1573 mi (360 × 2532 km). Incl. 65°. Total length, with rocket case, 80 in. (2 m). Explorer 1 carried 18 lb (8 kg) of instruments, designed to gather and transmit data on cosmic rays, meteorites and orbital temperatures. Confirmed the existence of belt of intense radiation beginning 600 mi (965 km) above the earth. Continued transmissions until May 23, 1958. Remained in orbit over 12 years, re-entering over S. Pacific after 58,000 revolutions on Mar 31, 1970. **Explorer 2** L. Mar 5, 1958 by Jupiter C from Cape Canaveral. Wt 31 lb (14 kg). Failed to orbit due to unsuccessful 4th-stage ignition. **Explorer 3** L. Mar 26, 1958 by Jupiter C from Cape Canaveral. Wt 31 lb (14 kg). Orbit 121 × 1746 mi (195 × 2810 km). Incl. 31°. Similar to Explorer 1, but with addition of small magnetic tape-recorder able to release 2 hrs of stored data on cosmic-ray bombardment in 5 secs as satellite passed over ground station. Transmitted until Jun 16, 1958; decayed Jun 28, 1958. **Explorer 4** L. Jul 26, 1958 by Jupiter C from Cape Canaveral. Wt 38 lb (17 kg). Orbit 163 × 1373 mi (262 × 2210 km). Incl. 50°. Mapped Project Argus radiation until Oct 6, 1958; decayed Oct 23, 1959. **Explorer 5** L. Aug 24, 1958 by Jupiter C from Cape Canaveral. Wt 38 lb (17 kg). Failed to orbit: upper stage fired in wrong direction, leading to collision of rocket and instrument section. **Explorer S-1** L. Jul 16, 1959 by Juno 2 from Cape Canaveral. Wt 92 lb (42 kg). Failed to orbit; destroyed by range-safety officer. The last of 5 successive failures (the others were 2 Vanguards and 2 Discoverers). **Explorer 6** L. Aug 7, 1959, by Thor–Able from Cape Canaveral. Wt 142 lb (64·4 kg). Incl. 47°. Similar to Pioneer 1, but used to send a

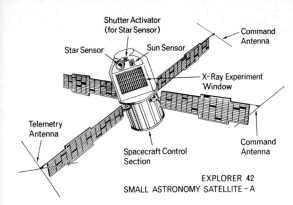

Shutter Activator
(for Star Sensor)

Command
Antenna

Star Sensor

Sun Sensor

X-Ray Experiment
Window

Telemetry
Antenna

Command
Antenna

Spacecraft Control
Section

EXPLORER 42
SMALL ASTRONOMY SATELLITE – A

'Paddlewheel' satellite, 26 in. (66 cm) dia., including 4 solar panels, or paddles, 20 in. (50 cm) square for recharging the batteries. Instruments measured behaviour of radio waves in ionosphere, mapped earth's magnetic field, cloud cover etc. It also sent back first photograph of earth. Transmitted until Oct 6, 1959; decayed Jul 1961. **Explorer 7** L. Oct 13, 1959, by Juno 2 from Cape Canaveral. Wt 92 lb (41 kg). Orbit 346 × 676 mi (557 × 1088 km). Incl. 50°. Returned data on earth's magnetic field and solar flares until Jul 24, 1961. **Explorer S-46** L. Mar 23, 1960 by Juno 2 from Cape Canaveral. Wt 35 lb (16 kg). Failed to orbit; upper stage apparently failed to ignite. **Explorer 8** L. Nov 3, 1960 by Juno 2 from Cape Canaveral. Wt 90 lb (41 kg). **Explorer 42, 48** Small Astronomy Satellites, which are building up a sky map of gamma rays—electromagnetic radiations which cannot be detected on earth because they are absorbed in its atmosphere. Explorer 42 was launched Dec 12, 1970 by Scout from Italy's equatorial platform at San Marco, wt 675 lb (306 kg), into 324 × 350-mi (521 × 563-km) orbit with 3° incl. Its data on X-ray sources and their location were followed up by Explorer 48, launched Nov 15, 1972 from San Marco, into a 275 × 393-mi (442 × 632-km) orbit with 1·9° incl. This 410-lb (186-kg) satellite was equipped with a spark-chamber gamma-ray telescope to study gamma rays and X-ray sources emanating from the galactic plane and the Crab Nebula. It is hoped that this series, to be completed with a third launch in 1975, will lead to some understanding of the dynamics of the Milky Way.

HEAO—ASTRONOMICAL OBSERVATORY

History 3 High Energy Astronomy Observatories are planned for launch between 1977 and 1979 to study some of the most intriguing mysteries of the universe. HEAO was originally planned as a £114.5 million ($275 million) project, with satellites 30 ft (9 m) long, 9 ft (2·7 m) dia., and wt 21,000 lb (9500 kg) in orbit. NASA budget cuts in 1973 resulted in launch delays and a reduction in satellite size to about 5000 lb (2270 kg), with single mission costs of £12.5 million ($30 million) instead of about £41.5 million ($100 million). Under the scaled-down programme, 3 spacecraft will be launched by Atlas/Centaur rockets, with additional launches from the Space Shuttle starting about 1980.

Objects: To study Pulsars and Neutron Stars,

HEAO studying pulsars, quasars, black holes etc. *TRW painting by Arthur Hill*

believed to be collapsed stars so closely packed that a spoonful of the centre would weigh a billion tons; Black Holes, believed to be the final stage in the collapse of a dying star; Quasars, remote objects emitting vast quantities of energy; Radio Galaxies, on the fringes of visibility, and emitting radio waves millions of times more powerful than normal spiral galaxies; and Supernovas, large stars whose final collapse is a cataclysmic event generating a violent explosion.

INTELSAT

INTELSAT (the International Telecommunications Satellite Consortium) is a partnership of 83 nations formed to establish a global commercial communications system. Its satellites have brought clear and reliable TV, telephone and other communications to countries and areas which had little hope of enjoying good reception before the age of the satellites. Intelsat itself is the prime example of the way in which a new technology both creates and satisfies new demands. It now carries more than two-thirds of all long-distance international communications.

History

It was established in August 1964 by 14 countries on the basis that each country invested in the satellite system in proportion to its expected use, and shared accordingly in the revenues. Since Intelsat I (the famous Early Bird) was launched the following April, 3 successive generations of more advanced satellites have been deployed, and a series of 'extended-capacity' Intelsat 4s is being considered.

Early Bird provided 240 high-quality voice circuits, and made transatlantic TV possible for the first time when placed in synchronous orbit over the Atlantic. Only 7 years later, the fourth-generation Intelsats (4s were in position by the end of 1972) were each able to carry an average of 5000 telephone calls, plus TV. From 5 TV transmissions a month in 1965 via Early Bird, the average was 100 a month by 1970. In 1972, with 225 satellite pathways available, there were 6790 transmission hours of TV, and 3725 satellite circuits were being leased on a full-time basis. During the 2 weeks of the 1972 Olympic Games at Munich, more than 1000 hours of TV were transmitted to 25 countries. The system played a major part in all 10 manned Apollo flights, both by providing support communications and in making world-wide TV coverage possible. The manned flights created an enormous demand for live TV coverage; and as quality improved and colour was added, there was a rapid growth in demand for satellite transmissions of news and sports events, both 'live' and in the form of edited and recorded packages, sent shortly after the event.

By 1962 it was clearly established that communications satellites were a commercial proposition. The setting up of Intelsat, and its management body, COMSAT (the Communications Satellite Corporation), quickly followed. Inevitably there was some international concern about the possibility that the US and the Soviet Union (*see* Molniya, page 164) would each develop a world monopoly of the commercial and operational aspects of their respective systems (though Molniya, highly elliptical, rather than synchronous, is still largely domestic).

Until February 12, 1973 Intelsat was managed on behalf of the member nations by Comsat;

this body is the US member of the Intelsat consortium, and a major investor in the satellites. Earth stations—65, in 49 countries, by the end of 1972—are needed to transmit and receive the flow of telephone calls, TV and other signals passed through the satellites, and these are usually owned by the countries in which they are located. Boosters and launch services for the satellites are brought from NASA at Cape Kennedy. Intelsat pays NASA for both the rockets and launch facilities.

Launch Chronology

Intelsat 1 (Early Bird) Dia. 28·4 in. (70 cm). Ht 23·25 in. (60 cm). Wt 150 lb (68 kg) at launch, 85 lb (38·5 kg) after apogee motor fire. Capacity 240 circuits or 1 TV Channel. Launcher: Thrust Augmented Delta. Early Bird, the only one in this series, was launched Apr 6, 1965; became operational over Atlantic at 325°E Jun 28, 1965. The world's first commercial communications satellite, it made transocean TV possible for the first time. Its antenna was focused N. of the Equator to service N. America and Europe. Design life was 18 months, but operated satisfactorily for $3\frac{1}{2}$ yrs before being placed in orbital reserve. **Intelsat 2** This series extended satellite coverage to two-thirds of the world, though only the last 2 achieved the planned 3-yr life. Wt 357 lb (162 kg) at launch, 190 lb (86 kg) after apogee motor fire. F-1, L. Oct 26, 1966, failed; F-2, L. Jan 11, 1967, operational for 2 yrs over Pacific. F-3, L. Mar 22, 1967, operational $3\frac{1}{2}$ yrs over Atlantic; F-4, L. Sep 27, 1967, operational $3\frac{1}{2}$ yrs over Pacific. **Intelsat 3** Capacity increased to 1200 circuits or 4 TV channels, with 5-yr design life (which was not achieved). Wt 647 lb (293 kg); 334 lb (151 kg) after apogee motor fire. F-1, L. Sep 18, 1968, failed; F-2, L. Dec 18, 1968, operational $1\frac{1}{2}$ yrs over Atlantic. F-3, L. Feb 5, 1969, operational for unstated period over Indian Ocean. F-4, L. May 21, 1969 operational 3 yrs over Pacific. F-5, L. Jul 25, 1969, failed. F-6, L. Jan 14, 1970, operational 2 yrs over Atlantic, now in reserve. F-7, L. Apr 22, 1970, operational $1\frac{1}{4}$ yrs over Atlantic. F-8, failed. **Intelsat 4** Capacity increased to 5000 circuits or 12 TV channels, with 7-yr design life. Wt 3120 lb (1414 kg); 1610 lb (730 kg) after apogee motor fire. F-1, held in storage. F-2, L. Jan 25, 1971, operational over Atlantic;

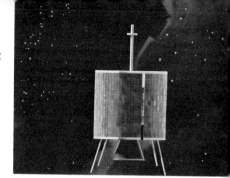
Intelsat 1: the history-making 'Early Bird'

Intelsat 2: twice the size of Intelsat 1

Intelsat 3: design life of 5 years not achieved

F-3, L. Dec 19, 1971, operational over Atlantic. F-4, L. Jan 22, 1972, operational over Pacific. (Assembled by British Aircraft Corporation at Bristol, F-4 was first commercial satellite assembled outside US, and provided TV link between US and China for President Nixon's visit to Peking.) F-5, L. Jun 13, 1972, operational over Indian Ocean. F-7, L. Aug 23, 1973, operational over Atlantic. F-6 & F-8 scheduled for launch in 1973/4. **Intelsat 4A:** First launch of 3 4A's will be in mid-1975. With launch wt 3240 lb (1470 kg), these will have nearly double the capacity of the Intelsat 4 series; they will be placed over the Atlantic, Pacific and Indian Oceans.

Summary The placing of Intelsat 4F–5 above the Indian Ocean, with an intended operational life of 7 years (and an orbital life, like the others in this series, of over a million years) brought the total of TV channels available to 60. By the end of 1972, there had been 17 Intelsat launches; 10 of the satellites were still operational; 4 had failed; 2 were partial failures, and Early Bird had been retired. The overall cost of the Intelsat 4 programme will be £97.8 million ($235 million); average cost of each satellite is £5.6 million ($13.5 million); and average cost of each launch and spacecraft is £12.3 million ($29.5 million). Thus, a launch failure represents a lost investment of £12.3 million ($29.5 million).

Synchronous Satellites It was in 1945 that Arthur C. Clarke, the space writer, suggested that a satellite placed 22,300 miles (35,680 km), from earth, would orbit at the precise speed required to appear to be hanging in space above one point on the planet's surface. Such a 'synchronous', or 'stationary' satellite would always be in position to relay radio, television and telephone signals, and no elaborate tracking devices would be needed. Compared with conventional underwater cables, satellite communications offered flexibility and limitless capacity. The first synchronous

In background, completed Intelsat 4; in foreground, another being assembled at Hughes Aircraft, El Segundo, Calif

satellite, Syncom 1, was successfully placed in orbit in February 1963, but its radio equipment failed to work. Syncom 2, launched in July 1963, was then placed in a 22,230-mile (35,567-km) orbit, and during the following 3 weeks manoeuvred into position over Brazil by the firing of jets from small, onboard hydrogen peroxide rockets. On September 13, Syncom 2 and Relay 1 were used to link Rio de Janeiro, New Jersey and Lagos, Nigeria, in a 3-continent conversation. But Syncom 2, not quite in the plane of the equator, appeared to describe a figure 8 as the earth turned beneath it. Syncom 3, weighing 85 lb (38·5 kg), was placed in true equatorial orbit, with no north–south swing, on August 19, 1964: perigee 22,164 miles (35,670 km); apogee 22,312 miles (35,700 km); period 1436·2 mins; incl. 0·1°. Drifting over the Pacific Ocean near the International Dateline on October 10, it telecast the opening-day ceremonies of the Olympic Games in Tokyo.

Launch Technique This can be illustrated by a description of the launch procedure for the proposed Extended Capacity Intelsat 4. The 3320 lb (1506 kg) satellite is mounted on the Atlas/Centaur launch vehicle, protected by a 10-ft (3-m) fibreglass nose fairing. Once clear of the earth's atmosphere, this is jettisoned during the first burn of the Centaur upper stage. Between them, the Atlas booster and the Centaur place the satellite in a 115 × 400-mile (185 × 644-km) parking orbit. After a 15-min. coast period, Centaur is again ignited. This provides the satellite with sufficient velocity so that, on separation from Centaur, it will coast to an altitude of 22,300 miles (35,680 km) above the earth's surface. Final synchronous equatorial orbit is achieved by firing the solid-fuelled apogee motor, which provides 34 secs of thrust at 12,500 lb (5670 kg). The altitude and speed of the spacecraft then matches the rotational speed of the earth, so that it remains over the same point. Small gas jets are then used so that the spacecraft drifts in orbit and is finally positioned accurately at any desired point on the equator. This process can take up to 2 months, and the spacecraft is not brought into operational use until this has been achieved. The jets are also used for 2 other vital manoeuvres. One is to orient the spacecraft so that it is correctly lined up with the earth, and its solar cells are in position to gather heat from the sun to charge its batteries. The other is to set the spacecraft spinning at the rate of 60 revolutions per minute; this stabilizes its position, on the same principle as a gyroscope; 40% of the spacecraft is 'despun' or contra-rotated, by means of a ball-bearing and power-transfer assembly, so that the highly directional antennae can be kept pointed towards the earth. The satellite carries 300 lb (136 kg) of hydrazine propellant for these manoeuvres; once this is exhausted, the spacecraft ceases to be operational, despite its orbital life of a million years. (When the Space Shuttle becomes operational, it should be possible to recover or refuel such satellites.)

Intelsat's Future Following a 6-week conference ending in May, 1971—described by the Chairman as 'very tedious'—it was decided that, since other nations were using the system at a faster rate than the Americans, Intelsat should no longer be wholly American-managed. As an interim measure, an Intelsat Secretary-General has been appointed, and has given Comsat a technical operations contract until 1978. In 1977 the Secretary-General will be replaced by a Director-General and an International Board of Directors for Intelsat. Any country will then be able to bid for technical operations contracts; the American Comsat firm may bid, but will no longer be guaranteed the job it has done since 1964. It is hoped this will resolve international frictions; and also make it easier to settle a parallel problem, involving international governments and airlines, as to whether Comsat should be allowed to establish satellite communications for air traffic control.

Any country belonging to the International Telecommunications Union can join Intelsat, and only 4 are not ITU members—China, N. Korea, N. Vietnam and E. Germany. The Soviet Union, which operates its own communications satellites, sends representatives to Intelsat meetings, but has not joined.

A request from the United Nations that Intelsat should provide free satellite channels for UN communications and international emergencies is being considered.

LUNAR ORBITER

History The second of 3 unmanned exploration projects, carried out in parallel with the 3 manned programmes aimed at getting men on the moon before 1970. Following the successful Project Ranger flights, which yielded the world's first TV pictures of the moon's surface, 5 Lunar Orbiters were launched within a year, starting on August 10, 1966 to help select Apollo landing sites in equatorial regions from 43°E to 56°W. Other objectives were to study variations in lunar gravity, radiation and micrometeoroid data. Orbiter 1 was placed in a 119 × 1160-mile (191 × 1867-km) lunar orbit, with 12° inclination. Its pictures, covering 2 million sq. miles (5·18 million sq. km), of the moon, included 16,000 sq. miles (41,440 sq. km) of potential Apollo landing areas. Perturbations in its orbit provided the first knowledge of what became known as 'mascons'— at least a dozen mass concentrations of materials, usually associated with the mare, which have a powerful gravitational effect on spacecraft remaining for lengthy periods in lunar orbit. All 5 Orbiters were immensely successful; it proved possible to manoeuvre them by earth commands into orbits descending as low as 25 miles (40 km). Objects as small as 3 ft (0·9 m) across were photographed, and their pictures provided the first lunar atlas including the far side, to be built up. The first 4 Orbiters provided between them experience

Checking the camera lens on the 5th Lunar Orbiter

of several thousand lunar orbits before each was deliberately crashed on to the surface with the last of its attitude control gas, to ensure that there was no radio frequency interference with later missions. Orbiter 4 provided the first pictures of the lunar south pole. Orbiter 5, launched on August 1, 1967, after it had completed its photography was retained as a target for NASA's Manned Spaceflight Network while it had sufficient fuel, until its final controlled impact on January 31, 1968. By then the third unmanned lunar exploration project, Surveyor, was also nearing completion.

Spacecraft Description A truncated cone structure. Lunar Orbiters were folded for launch. When deployed, the 4 windmill-like solar panels and antennae provided a maximum span of 18½ ft (5·6 m), and 5 ft 6 in. (1·6 m) ht. Total wt of 860 lb (390 kg) included a photographic laboratory weighing only 145 lb (65·8 kg), but carrying 2 cameras for wide-angle and telephoto coverage, film processing and photo readout (scanning) systems. These viewed the moon through a quartz window protected by a mechanical flap. Launcher: Atlas–Agena D.

MARINER

History One of NASA's 3 planetary exploration programmes. It has complemented and leap-frogged Pioneer, and built up the technology for the Viking landings to be attempted on Mars in 1975. The first 9 launches, spread over 10 years, included 3 intended for Venus and 6 for Mars; their success has enabled the tenth to be aimed at Mercury, passing Venus on the way, with the following pair intended to fly past Jupiter and Saturn. The latter replaced the more expensive 'Grand Tour' missions intended to use the period 1977–79 when all 5 outer planets will be lined up in such a way that their gravity could be used to swing spacecraft past each in turn; this occurs only once in 180 years. The revised plan is costing only £133.3 million ($320 million) compared with the Grand Tour's £375 million ($900 million). The 2-yearly launch 'windows' (when the earth and Mars come within about 35 million miles (56 million km) of each other) were used by Mariner 4 in 1967 to obtain man's first close look at another planet; and by Mariners 6 and 7 in 1969 to follow up with much better pictures. In 1971 Mariner 9 became man's first planetary orbiter, and provided us with a complete map of the Red Planet. The secrets of Venus, still largely concealed beneath her dense cloud cover, despite Russia's landing attempts, will be explored by both Mariner and Pioneer missions during the remainder of this decade. Whatever their success, the place of Mariner in space history is secure.

Spacecraft Description The early Mariners, 9 ft 11 in. (3·04 m) long, and 5 ft (1·52 m) across the base, were very similar to the Ranger space-craft used for impacting on the moon; a tubular centre was attached to an hexagonal base, from which a dish antenna and solar cell panels were extended. The weight of Mariners 3 and 4, as a result of a more powerful launcher, was increased from 446 lb (202 kg) to 575 lb (260 kg). An octagonal magnesium centrebody had 4 rectangular solar cell panels to power its computer and sequencer, TV camera, cosmic-ray telescope etc, and a hydrazine-fuelled main engine. Mariners 6 and 7, twice as large again, 11 ft (3·35 m) high, had an 8-sided magnesium framework with 8 compartments containing electronics, TV assembly etc; and 4 rectangular solar panels 84 in. (213 cm) long, with attitude control jets on the tips of the panels. Each had 2 TV cameras (wide and narrow angle) mounted on a rotating platform, able to resolve objects down to 900 ft (275 m). Mariners 8 and 9 retained the same basic design, but the need for a 300-lb (136-kg) thrust retro-engine to inject them into Mars orbit again increased total weight to 2272 lb (1031 kg); this included 1000 lb (454 kg) of fuel. The narrow-angle TV camera could resolve features down to 300 ft (100 m). Other instruments for investigating the atmosphere and surface included an infra-red radiometer, ultra-violet spectrometer, and an infra-red interferometer spectrometer.

Mariner 1 L. July 22, 1962 by Atlas–Agena B from Cape Kennedy. Wt 446 lb (202 kg). This first attempt at a Venus fly-by failed because of an error in the flight guidance equation; the rocket went off course immediately after launch, and had to be blown up. The object had been to obtain details of the Venusian atmosphere, cloud cover, magnetic field etc.

Mariner 2 L. August 27, 1962 by Atlas–Agena B from Cape Kennedy. Wt 447 lb (202 kg). The first successful planetary fly-by, it was fired into a Venusian trajectory from earth-parking orbit; after a 109-day journey it flew past the planet at a distance of 21,648 miles (34,830 km), providing 35 mins of instrument scanning. Surface temperatures registered at 428°C, above the melting point of lead, and far higher than expected. The atmosphere appeared to contain no water vapour. The cloud layer was unbroken, with one spot near the southern end of the terminator 11°C cooler than the rest, possibly due to a mountain range. It was also established that, unlike the earth, Venus did not have a strong magnetic field and radiation belt.

Mariner 3 L. November 5, 1964 by Atlas–Agena D from Cape Kennedy. Wt 575 lb (261 kg). Intended to take 21 TV pictures as it passed Mars at a distance of 8600 miles (13,840 km), but failed to achieve the necessary speed of 25,661 mph (41,228 kph), when fired from earth-parking orbit. Although it went into solar orbit, Mars was missed by a wide margin.

Mariner 4 L. November 28, 1964 by Atlas–Agena D from Cape Kennedy. Wt 575 lb (261 kg). Following Mariner 3's failure, launch of the second of the pair of vehicles was delayed till the end of the Martian window. Injection was successful, and problems resulting from the instruments locking on to the wrong stars were overcome; 2 days after launch, Canopus was acquired, and after 228 days and a flight of 325 million miles (523 million km), Mars was passed at a distance of 6116 miles (9844 km), on July 14, 1965. During the next 10 days 21 TV pictures, and 22 lines of a 22nd photograph, were received at NASA's Jet Propulsion Laboratory at Pasadena, California. Man's first close-range pictures of another planet showed that Mars was heavily cratered, more moon-like than earth-like, very dry, with no trace of surface water, and certainly not possessing any of the canals theorized by astronomers. Although the possibility of some form of life could not be ruled out, the very thin atmosphere, coupled with the evidence that there might never have been enough water for oceans or streams, made any advanced life seem most unlikely. Long after passing Mars, Mariner 4, in solar orbit, provided convincing evidence that such vehicles could be operated for many years; $2\frac{1}{2}$ years later, 56 million miles (90 million km) from earth, a JPL command again turned on the TV equipment, and fired the spacecraft engine for 70 secs.

Mariner 5 L. June 14, 1967 by Atlas–Agena D from Cape Kennedy. Wt 540 lb (245 kg). Originally the back-up vehicle for Mariner 4, this was modified for flight towards the sun and Venus, instead of away from the sun to Mars. Solar panels were reversed and reduced in size, and a thermal shield added. A flight of 217 million miles (349 million km) resulted in Mariner 5 passing only 2480 miles (3990 km) ahead of Venus in its orbit around the sun, on October 19, 1967. Using more advanced instruments, surface temperatures of

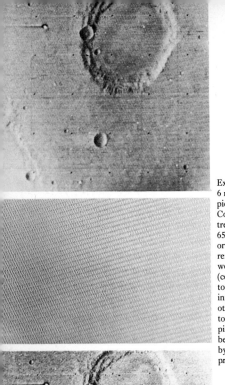

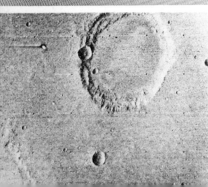

Example of Mariner 6 near-encounter picture of Mars. Computer treatment of the 658,240 elements in original picture removes the 'basket weave' pattern (central picture) due to electronic interference, plus other interference, to results in last picture. Pictures can be further improved by additional processing about 267°C were recorded; measurements of the magnetic field ranged between zero and 1/300th of earth's; an electrified ionosphere was identified at the top of the atmosphere.

Mariner 6 and 7   L. February 24 and March 27, 1969, by Atlas–Centaur from Cape Kennedy. Wt 910 lb (413 kg). Intended to study the atmosphere and surface of Mars as part of the search for extraterrestrial life, and to develop technology for later Mars missions, these flights were immensely successful. The author, watching the 201 TV pictures flowing back to JPL at Pasadena, found it even more exciting and dramatic than covering the first Apollo moonlanding a few days earlier. They passed Mars at distances of 2120 and 2190 miles (3412 and 3524 km). Mariner 6 had flown 241 million miles (387·8 million km) in 156 days, to arrive on July 31, for encounter at 59·5 million miles (95·7 million km), about 5½ light mins, from earth. Mariner 7 flew 197 million miles (316·9 million km) in 130 days for encounter on August 5 at 61·8 million miles (99·4 million km). Mariner 7 was probably struck by a small meteoroid a few days before arriving; after loss of signal, commands sent ordering it to switch antennae were successful both in restoring communications and establishing that it had been damaged, losing some of its telemetry channels; a slight velocity change caused it to arrive 10 secs late. The spacecraft began sending back far-encounter pictures from distances of up to 700,000 miles (1,126,540 km); but greatest interest naturally lay in the 24 near-encounter pictures sent back by Mariner 6 during its 68 mins of closest approach, and in the 33 near-encounter pictures provided by Mariner 7's 74 mins of closest approach. Mariner 6, concentrating on the equatorial region, dramatically established

65

that Nix Olympica, at first thought to be a gigantic crater, was a 15-mile (24-km) high volcano, with a 40-mile (64-km) wide crater at the top. Mariner 7, concentrating on the southern hemisphere and part of the south polar ice cap, confirmed that this was largely solid carbon dioxide (dry ice), with perhaps a little water content. Mars emerged at the end of the fly-bys as heavily cratered, with a thin atmosphere consisting of at least 98% carbon dioxide, its craters differing from those on the moon as a result of being worn down by winds and dust. It seemed that any advanced form of life could be ruled out, but one scientist speculated on the possibility of some form of life evolving which by-passed the need for liquid water. One of the Mariner 7 pictures showed a minute, potato-shaped speck which proved to be one of the two Martian moons, Phobos; but full details of the moons and of the Red Planet itself were to be finally established by Mariner 9 during the next launch window. Cost of this twin-mission was £61.6 million ($148 million).

Mariner 8   L. May 8, 1971 by Atlas–Centaur from Cape Kennedy. Wt 2272 lb (1031 kg). Intended to be the first of a pair of Martian orbiters, but an autopilot fault sent the second stage off course and it fell into the Atlantic 900 miles (1450 km) SE of Cape Kennedy.

Mariner 9   L. May 30, 1971 by Atlas–Centaur from Cape Kennedy. Wt 2272 lb (1031 kg). Intended to map 70% of Mars during 90 days in orbit around it, Mariner remained operational 349 days before it was shut down on October 27 following exhaustion of its attitude control nitrogen gas; by then it had circled the Red Planet 698 times, mapped the whole of it, and transmitted 7329 TV pictures, including detailed photographs of both Phobos and Deimos. Following the loss of Mariner 8, plans were revised so that Mariner 9 could cover both missions. The spacecraft arrived at Mars on November 13, 1971, at the end of a 167-day flight covering 248 million miles (397 million km). A 15-min firing of its 300 lb (136 kg) thrust liquid propellant engine reduced the approach speed, relative to Mars, from 11,185 mph (18,000 kph) to 7830 mph (12,500 kph), and placed it (after a later trim manoeuvre) in a 12-hr Martian orbit, $10,650 \times 852$ miles ($17,140 \times 1387$ km). It thus became the first man-made object to orbit another planet. (Russia's Mars 2 and 3 followed later in 1971.) The braking burn reduced the spacecraft's weight to 1300 lb (590 kg). As the spacecraft was approaching Mars in mid-November, it took 3 series of pictures of a violent dust-storm which astronomers had been watching envelop the entire Martian globe during a 2-month period. While this delayed Mariner 9's mapping sequences for 6 weeks, it provided a unique opportunity for its instruments to peer down into the most extensive dust storm to occur on Mars since 1924. It reached an altitude of 30–35 miles (50–60 km). Only the bright, waning ice cap at the south pole, and 4 dark mountain peaks (one of them Nix Olympica) were visible through the haze, which had the effect of cooling the surface, and warming the atmosphere—an indication to earth scientists of the effects of increasing pollution of the earth's atmosphere. When the dust storm subsided Mariner 9 was able to maintain an instrument surveillance of the changing seasons below for more than half a Martian year. By the end of its mission Mars was known to be a geologically active planet, different from both earth and moon, with volcanic mountains and calderas (craters) larger than any on earth; there is a vast equatorial crevasse which would dwarf America's Grand Canyon, 2500 miles (4000 km) long, and plunging to a depth of 20,000 ft (6096 m). The 'Martian canals' were an illusion, yet this gigantic rift was never suspected, showing up on earth-based telescopes only as dark markings. One theory is that the dark trough is warmed by the sun at one end, while it is still dark at the other, with the effect that violent winds are set up each day. Contrary to earlier conclusions, it is now thought that free-flowing water may have existed on Mars at one time; and that dust storms and cloudiness account for much of the variability of appearance that has puzzled astronomers for centuries. Because of a previously unknown gravity-field variation in Mars' equatorial plane, Mariner's orbital period was found to be too short in relation to its earth tracking stations; so, after the dust storm cleared, its engine was fired to raise the periapsis, or low point, to 1025 miles (1625 km), to co-ordinate the orbit with the Goldstone, California, station. With the Martian surface clear at last, the mapping cameras looked down on a shrinking south polar cap; sinuous channels which appeared to be dried-up river beds cut by water; chaotic, bouldery terrain first glimpsed by Mariners 6 and 7; and huge impact craters, their floors covered with wind-blown dunes. Surface temperatures ranged from 81°F on the equator to −189°F at the poles; the north pole was much colder than earth's coldest spot, which is Antarctica, at −125°F. Several localized dust storms were seen after the main storm cleared. Variable cloud patterns were observed, mainly in the north, but also over large volcanoes, and were believed to contain water ice; though if large quantities of water exist, they seem certain to be locked in the permanent polar ice

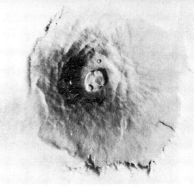

Nix Olympica, shown by Mariner 9 to be a mountain, not, as astronomers had thought, a huge crater

caps. Atmospheric winds were measured up to 115 mph (185 kph). Nix Olympica, 300 miles (500 km), across its base, is the Red Planet's highest spot, the peak reaching at least 10½ miles (17 km) above the surrounding plain, making it far higher than Everest on a planet half the size of earth. The tiny Martian moons were both studied: Deimos, orbiting at 12,471 miles (20,070 km), has a 10-mile (16-km) dia. equator, and is 6 miles (9·6 km) from north to south. Phobos, orbiting at only 3720 miles (5986 km), has a 17-mile (27·3-km) dia. equator and is 12 miles (19·3 km) from north to south. Both moons are heavily cratered, apparently from meteorite impacts; gravity is so low on Phobos that a man could throw a cricket ball into orbit around it. From April 2 to June 4, 1972, Mariner's instruments were turned off while its orbit took it into Mars' shadow during each twice-a-day revolution. After they had been successfully turned on again it became possible to study the north polar region, and to look for potential landing sites for the Viking 1975 project. The last of 45,960 commands to Mariner 9 was to turn off its radio transmitter; it is expected to remain in Martian orbit for at least 50 years. The costs of Mariners 8 and 9 totalled £56.7 million ($136.4 million).

Mariner 10 L. November 3, 1973 by Atlas/Centaur from Cape Kennedy. Wt 1108 lb (503 kg). It is due to pass Venus in February 1974, at a distance of 3300 miles (5300 km) to make further studies of its dense cloud blanket, and of the recent discovery that the top of the clouds oscillate, or move up and down, more than 0·6 miles (1 km), in a constant wave motion. It will then use Venusian gravity to deflect its trajectory to pass within 625 miles (1000 km) of Mercury, in late March 1974. The heliocentric trajectory, resulting from the near-Mercury fly-by, should ensure a second and third encounter in September 1974 and March 1975, providing further opportunities for the TV camera to study Mercury's landmarks, possible atmosphere and other features. This mission will cost up to £50 million ($120 million).

Mariners 11 and 12, weighing 1600 lb (700 kg), are due to be launched in August and September 1977 to pass Jupiter $1\frac{1}{2}$ years later, and Saturn after $3\frac{1}{2}$ years. They will be following up the Pioneer flights to the outer planets. Cost of M. 10–12 flights is estimated at £133.3 million ($320 million).

Planet Mercury

Of the 9 planets, this is much the nearest to the sun—36 million miles (58 million km), compared with earth's 93 million miles (150 million km). With a diameter of 3100 miles (4989 km), compared with earth's 8000 miles (12,875 km), it is the smallest. Radar observations have recently established that the long-held view that Mercury always keeps the same face to the sun is wrong; in fact it rotates $1\frac{1}{2}$ times during each revolution, and has only that number of days in its year, which lasts 88 earth days. Because it is so close to the sun, Mercury's surface temperature is expected to be intolerably hot for man, without water, and with little or no atmosphere. It is expected to be heavily cratered; but, prior to the Pioneer 10 mission, Mercury had received little study, although its proximity to the sun makes it one of the most important planets in man's quest for data on the origin and history of the solar system.

MILITARY SATELLITES

History The wide variety of military, reconnaissance or 'spy' satellites now being launched by both the United States and the Soviet Union, can be broadly divided into 5 categories:

1 Early Warning
2 Nuclear Explosion Detection
3 Photo Surveillance
4 Electronic Surveillance
5 Communication and Navigation.

This entry seeks to lift at least a corner of the veil of secrecy surrounding US military activities in space, by co-ordinating the very limited amount of information officially announced, with expert deductions made from the orbits and weights worked out by independent observers in Britain and the US. By mid-1973 America's military launches numbered around 310, compared with about 430 by Russia; but many US Defense Department launches are classified as civil rather than military; these are announced launches of small research satellites like Radcat and Radsat (L.

October 2, 1972, from Vandenberg) as radar calibration targets, and to measure radiation effects on the lifetimes of instruments.

In the same way, satellites providing telephone communications can hardly be classified as 'military' in the strict sense. Since it is not so easy to differentiate in this way with Soviet Cosmos launches, this may account, to a limited extent, for the apparently much higher Russian launch rate; but it will be noted that Russia's total of both military and civil launchings overtook America's in 1967, and has remained more than double the American figure ever since. This has led to fears among some of America's military space experts about the vulnerability of their much thinner space systems. They fear that one day Russia might attempt to 'blind' American defences by intercepting and destroying her early-warning satellites, and by putting her navigation and communication satellites out of action by attacking them with laser beams. To counter this possibility, much work is being done to increase the survivability of America's military satellites by shielding them against laser beams, and by keeping a close watch with the IMEWS system, not only on Russian launches, but on the performance and behaviour of the rival satellites when they are in orbit; and by providing multiple communications paths, in various frequencies, by means of multi-purpose systems—'satellite buses', as they are called. The result of this emerges in the entries below: having branched out into various 'families', the classes of satellites are now beginning to merge again.

It should be remembered that the 'space race' began, not so much because of the international prestige involved, but because Russia and America both feared that the other side might achieve complete military dominance if it became the first to master space technology. By 1960 the US Air Force's Ballistic Missile Division at Patrick Air Force Base, 20 miles (32 km) south of Cape Kennedy, had 14 major programmes aimed at 'searching out' the possible advantages of space activity. The design and use of military satellites was far advanced long before the first 2 Sputniks were orbited in 1957; and this section should be read in conjunction with the DISCOVERER and TIROS entries. Tiros 1 was the first US reconnaissance satellite to be successfully orbited; but, by then it had already been decided that BMEWS, consisting of 3 ground-based radar stations at Thule in Greenland, Clear in Alaska, and Fylingdales in England, providing America with 15 minutes', and Britain 4 minutes' warning if enemy missiles rose above the horizon, were not sufficient. Projects Midas (Missile Defence Alarm System) and Samos (Satellite and Missile Observation System) were the first efforts to use space to double the warning time.

Finally, a note on the future. When NASA has completed development of the Space Shuttle (details of which are given in the *Observer's Book of Manned Spaceflight*) it is expected that in the 1980s and 1990s this civil system will also be used for an average of 20 military launches a year. Since the Space Shuttle itself will only be capable of earth-orbit operations at altitudes of about 100 miles (160 km), it will, when it first comes into operation, have to carry up, as part of its payload, Centaur and Agena upper stages, which will then boost the military satellites to their synchronous and other much higher orbits. Later, the plan is to develop a reusable 'space tug', which will be carried into orbit, and refuelled by the shuttle, so that the tug can then both place synchronous satellites in orbit, and recover and refurbish those already there.

Early Warning Satellites

Midas   Equipped with highly sophisticated infra-red sensors capable of picking up the exhaust heat from a ballistic missile as it leaves the ground, this was originally intended to increase from 15 to 30 mins the advance warning for the USA of a missile attack, provided by BMEWS. The plan was to have 12 to 15 satellites in polar orbits. The 'readout stations', built to receive their information, and feed it into the BMEWS system, included one at RAF Kirkbride, Cumberland.

Midas 1 (L. February 25, 1960) failed to reach orbit as a result of a faulty stage separation; Midas 2 (L. May 24, 1960) achieved orbit, but telemetry failure prevented transmission of infra-red data. Midas 3 (L. July 12, 1961) was successfully launched from Vandenberg into a circular polar orbit of 2130 miles (3428 km), and 91° incl. Orbital life was 100,000 years; though it was known to be fully operational, there is no record of how long it lasted. Midas 4 (L. October 21, 1961) and Midas 6 (L. May 9, 1963) aroused an international furore, when details of an experiment called Project West Ford became known. This was to eject a 77-lb (35-kg) canister, containing 350 million hair-like, copper dipoles, $\frac{7}{10}$ in. ($\frac{18}{25}$ cm) long. The idea was that, after separation, the spinning canister should slowly dispense the dipoles in an orbital belt 2000 miles (3220 km) high, 5 miles (8 km) wide and 25 miles (40 km) deep, to test whether they would act as passive reflectors for relaying military communications. For over a year the project was violently attacked by the world's scientists, particularly in Britain

74

and Russia, because they felt the dipoles might interfere with astronomical observations, especially with radar telescopes. Professor Keldysh, President of the Soviet Academy of Sciences, said the experiment could result in 'serious contamination of near-terrestrial space and greatly hamper both manned spaceflights and astronomical observations'. Midas 4 successfully ejected its canister, but the dipoles failed to disperse; despite the protests, the US Air Force insisted on repeating the experiment with Midas 6, which was said to be successful; but after that, talk of dipoles was dropped. Apart from the dipole incident, Midas 4 was credited with detecting the launch of a Titan missile from Cape Kennedy 90 secs after lift-off. This incident inevitably led to increased secrecy being applied to military space experiments; but the space logs identified 9 Midas launches in circular 2000-mile (3220-km) polar orbits, up to October 5, 1966, carried out by Atlas–Agena D rockets.

IMEWS Integrated Missile Early Warning Satellite, a USAF project now operational as successor to Midas. From a geostationary orbit, it employs an infra-red 'telescope' to detect exhaust plume emissions from missiles as soon as they are launched. Immediate warning of hostile launches is believed to be transmitted to ground stations in Guam and Woomera in Australia, and from there via military communications satellites to NORAD (North American Defense Command headquarters) at Colorado Springs. In addition to the early warning role, such satellites can obviously monitor test launches, and are believed to have provided immediate information when Russia tested missiles carrying 3 separate warheads which spread out to make interception more difficult. They may already have taken over the early warning role of the Vela satellites. IMEWS 1, L. November 6, 1970, by Titan 3C from Cape Kennedy, wt 1800 lb (820 kg) was not completely successful. After being placed in a 16,200 × 22,400-mile (26,070 × 36,050-km) orbit at 7·8° incl., for checkout of the systems over the USA, fuel was exhausted before it could be moved into position to observe missile tests in China and firings along Russia's Pacific test range. IMEWS 2, L. May 5, 1971, from Cape Kennedy into a 26,175 × 22,152-mile (42,124 × 35,651-km) orbit with 0·87° incl., was successfully placed over the Indian Ocean; shortly afterwards the US Senate Armed Services' Committee was told there was a satellite 'capable of immediately reporting ICBM launches from the Soviet-Sino area.' IMEWS 3, L. March 1, 1972, was stationed in a 26,140 × 22,345-mile (42,067 × 35,962-km) orbit, with 0·2° incl., over the Panama Canal, where it can view both the Atlantic and E. Pacific oceans to detect a sub-marine-launched missile attack. It seems likely that IMEWS, powered by cruciform solar panels spanning about 23 ft (7 m), have an operational life of over 5 years; unless removed by future space tug operations, they will remain in orbit for around 1 million years.

Calsphere/Thorburner One of the most mysterious series, but possibly concerned with establishing a US ability to intercept and destroy enemy satellites. 3 tiny satellites, 2 later identified as Calspheres 1 and 2, were launched in a joint USAF/USN operation, on October 6, 1964, into 650 × 670-mile (1046 × 1078-km) orbits at 90° incl., and a successful interception manoeuvre is believed to have been carried out. Thorburner 2 launches (named after the rocket combination) have taken place twice a year into identical orbits since 1969; the fourth Thorburner 2, L. February 17, 1971, placed Calspheres 3, 4 and 5 into precisely similar orbits. The US Defense Department has said that while it has not matched Russia's satellite-intercept capability, it could be developed if required.

Nuclear Explosion Detection

Vela/ERS Launched in pairs, and planned before America's first satellite had been launched, Vela's main task is to detect and identify nuclear explosions in space. They will provide instant warning of any violation of the 1963 treaty which prohibits the testing of nuclear weapons either in the atmosphere or distant space. The initial pair, launched by an Agena D, on October 17, 1963, were manoeuvred by onboard rocket motors into circular 67,000-mile (107,825-km) orbits on opposite sides of the earth, well beyond the Van Allen radiation belts. Velas 3 and 4 were launched on July 17, 1964; and Velas 5 and 6 on July 20, 1965, by Atlas–Agena D launchers. Uprated Velas 7 and 8 were launched on April 28, 1967, and uprated Velas 9 and 10 on May 23, 1969, by Titan 3C launchers. Advanced Velas 11 and 12 were launched on April 8, 1970, also by Titan 3C. Similar orbits were employed in each case; transmission life of these satellites, which nowadays also includes solar flare and other observations, is probably about 3 years. Orbital life of the satellites is in all cases estimated at 1 million years.

Spacecraft Description The original Vela was 20-sided, 56 in. (1·42 m) wide, wt 510 lb (231 kg). Its 18 detectors could identify X-ray, gamma ray and neutron emissions and would have detected nuclear explosions as far away as Mars and Venus. The latest Velas are 26-sided polyhedrons, weighing 571 lb (263 kg), after burnout of the onboard solid-propellant apogee motor. Approx. 22,500 solar cells cover 24 sides to generate 120 W of electric power; the satellites are continuously oriented to look down into the earth's atmosphere from opposite sides. Vela launches invariably include piggy-back auxiliary payloads, such as ERS (Environmental Research Satellites). These range in weight from 1½–44 lb (0·7–20 kg), and usually carry a single scientific or engineering research experiment. 29 ERS launches had been made by the end of 1972.

Photo Surveillance

Samos A USAF programme, providing operational versions of the satellites first developed by the early Discoverers, and able to photograph all parts of the world from polar orbits, tape-recording TV pictures while over potentially hostile territory and transmitting them when passing over US territory; able to supplement and improve on these pictures by periodically dropping off capsules containing film.

Samos 1, L. October 11, 1960 by Atlas–Agena A from Point Arguello, wt 4100 lb (1860 kg), failed to achieve orbit. Samos 2, L. January 31, 1961, into a 344 × 295-mile (554 × 475-km) orbit, and 97° incl., successfully returned experimental data. Samos 3, L. September 9, 1961, exploded on the launchpad. Subsequent launches came after the decision that details of military satellites should be classified; but the series, operational since 1963, progressed through steadily more advanced versions. These are placed in polar orbits with perigee, or low points, of less than 100 miles (161 km), and on occasions specially launched to take a close look at points of interest picked out by routine surveillance satellites in much higher orbits. There is some evidence that Samos 87 was launched on March 17, 1972, by a Titan 3B/Agena D booster from Vandenberg into an 81 × 254-mile (131 × 409-km) orbit with 110° incl. It weighed about 6614 lb (3000 kg), was a 26-ft (8-m) long cylinder with 4·8 ft (1·5 m) dia.; typically, it remained in orbit 25 days.

Big Bird A huge USAF multi-function satellite, designed to perform both the 'search-and-find' and 'close-look' functions which required 2 different spacecraft until it came into operation in 1970. Also known as LASP, for Low Altitude Surveillance Platform. Weighing over 10 tons (10,160 kg) in orbit, it probably consists of a modified Agena rocket casing, 50 ft (15·2 m) long and 10 ft (3·05 m) dia., fitted with a high-resolution Perkin–Elmer camera capable of identifying objects as small as 1 ft (0·3 m) across from heights of more than 100 miles (160 km). Operational techniques are similar to those employed by the ERTS satellites: they are placed in sun-synchronous orbits so that they pass regularly over the targets at the same time of day. A series of pictures with identical sun angles is thus obtained, and changes occurring, such as the construction of new missile sites, and the number and types of missile being installed, are easily read. Film is processed on board, then scanned by a laser device, and converted into electronic signals transmitted back to earth to at least 7 receiving stations at the USAF's global bases. Drag encountered by such a large vehicle at such low altitude would mean that it would re-enter in 7–10 days; the Agena engine is therefore fired periodically to raise the orbit and extend its life. Big Bird is almost certainly capable of carrying out some Elint, or electromagnetic surveillance, as well, and sometimes carries with it a small 132-lb (60-kg) piggy-back capsule, placed in a higher orbit for such 'ferret' operations.

Big Bird 1, L. June 15, 1971, from Vandenberg, the first known launch by a Titan 3D/Agena, with wt of 25,130 lb (11,400 kg), was placed in a 70 × 115-mile (114 × 186-km) orbit, with 96° incl., and decayed after 52 days. By autumn 1973, 4 had been launched into similar but slightly higher orbits, and lifetime had been extended to 90 days.

Project 1010 Code-name for even more advanced satellites than Big Bird. Expected to be operational about 1977, they will be placed in geo-stationary orbits, and provide continuous, live TV coverage of any desired target; they will do this by passing their pictures back to earth via Data Relay Satellites.

Electronic Surveillance

Elint A general name for electronic intelligence, or 'ferret' satellites. Little is known about their numbers or effectiveness. They are usually launched into higher orbits—about 310 miles (500 km)—than photographic satellites. They record electromagnetic radiations being transmitted from areas of military activity, and replay them to ground stations for study and identification. By the end of 1972 nearly 30 Elint satellites had been identified, more than 20 having been launched with a photo satellite in lower orbit. The 'radar signatures', or characteristics, such as pulse repetition frequency, pulse width, transmitter frequency, modulation, and so on, enable the likely function and method of operation of a particular centre to be identified; the numbers and types of electronic systems at the site, and subsequent changes will give a valuable indication of its purpose and capability. Electronic Warfare, Electronic Countermeasures, and Electronic Counter Countermeasures, carried on at present by a wide variety of aircraft and ground-based stations, are likely to be increasingly taken over by satellites in the next decade. The ability to intercept and decode an enemy's satellite and other communications, and to interfere with them by rival satellite activities, is expected to be a decisive factor in any large-scale future hostilities.

Communication and Navigation

History A military satellite operating in synchronous orbit (where it revolves at the same speed as the earth, and therefore remains over the same point on earth), at an altitude of 22,300 miles (35,890 km), overlooks 63 million sq. miles (163 million sq. km) of the earth's surface, compared with an aircraft 5 miles (8 km) high, overlooking 110,000 sq. miles (284,900 sq. km). The development of these satellites, therefore, for military communications and navigation has formed a major part of the US space effort. There is room here only for a summary; and for clarity they are dealt with separately.

Communications The world's first communication satellite, the 21st in the world log to be orbited, was Score, L. December 18, 1958 by Atlas B from Cape Canaveral, into a 115 × 914-mile (185 × 1470-km) orbit and 32° incl. Wt 154 lb (70 kg). It transmitted taped messages for 13 days, and re-entered 34 days later. Next came Courier, L. October 4, 1960, the first active-repeater

Comsat, which operated for 17 days. Lack of rocket power, together with political and economic argument, delayed further developments until June 16, 1966, when a Titan 3C successfully orbited 8 satellites, including the first 7 of the Initial Defence Satellite Communications System. 26-sided polygons, 34 in. (86 cm) dia., covered with solar cells, and with no moving parts each weighed 100 lb (45 kg). Dispensed over a period of 6 hrs at slightly different orbital velocities to give them global coverage, they were placed just below synchronous altitude, at 21,074 miles (33,915 km). Drifting about 30° relative to earth, each stayed in view of an equatorial station for 4½ days, so that even if one malfunctioned there was always another drifting into position. Spin-stabilized, with a service life of 3 years, their electronic components were programmed to shut off automatically at the end of 6 years. The system, totalling 26 satellites, was completed by 3 more Titan 3C launches, the last on June 13, 1968. The satellites were capable of linking ground points 10,000 miles (16,090 km) apart, and from 1967 provided a South Vietnam–Hawaii–Washington link for transmitting, among other things, high quality reconnaissance photographs. Improved versions weighing 535 lb (242 kg) at launch, 285 lb (129 kg) after onboard rocket motors had manoeuvred them into stationary, equatorial orbits were NATO 1, L. March 20, 1970, and NATO 2, L. February 2, 1971; with 5-year operational life, they provide regular communications between operational ground stations. Skynet 1 and 2, somewhat similar, were launched for Britain on November 21, 1969 and August 19, 1970, to provide communications between Britain and her ships and military land bases overseas. Skynet 1 was successfully manoeuvred into position over the Indian Ocean; Skynet 2's apogee motor failed and had to be abandoned in its transfer orbit of 168 × 22,410 miles (270 × 36,065 km). As the 5-year life of the first IDSCS satellites was nearing its end, launch of more advanced replacements began. DSCS 1 and 2, L. jointly, November 3, 1971, by Titan 3C from Cape Kennedy. Wt 1150 lb (520 kg) cylinders, 9 ft (2·7 m) dia., 6 ft (1·8 m) tall. Their job in stationary orbit was to expend communications between US military installations around the world.

Meanwhile, development of tactical communications satellites, needing much greater onboard power, so that they could be received by small, low-powered ground terminals carried by ships, tanks, jeeps and aircraft, were also under development. LES 5 (Lincoln Experiment Satellite), L. July 1, 1967, from Cape Kennedy, wt 225 lb

Fleet Satellite Communication System (FLTSATCOM) for US Navy

(102 kg), as part of a 6-satellite payload (which included IDSCS 16, 17 and 18) was America's first. 2 days after it had been manoeuvred into a 20,729-mile (33,360-km), near-synchronous orbit, the first satellite communications between US aircraft, a US Navy submarine and surface vessel, and Army ground units, had been carried out. Tacsat 1, L. February 9, 1969, 25 ft (7·6 m) tall, 9 ft (2·7 m) dia., wt 1600 lb (725 kg), gyro-stat-stabilized so that the antennae and tele-scopes could be continuously pointed while the major part of the satellite spun within them, was a follow-up project; its object was to communicate with tiny land-based receivers only 1 ft (0·3 m) in diameter.

Navigation The first navigation satellite, Tran-sit, was developed primarily to provide Polaris missile submarines with the ability to fix their positions within one-tenth of a mile; it soon became evident that all-weather navigation could be provided in this way for all types of shipping.

Transit 2A, launched 1960, had a 2-year life

After an initial launch failure, Transit 1B, L. April 13, 1960, wt 265 lb (120 kg), into a 232 × 463-mile (373 × 745-km) orbit at 51° incl., transmitted information including time signals for 3 months which enabled navigators to fix their position. By 1968, 3 Transit series, totalling 23 satellites, had been placed in circular orbits of about 500 miles (805 km); in mid-1972, 5 were operational, 2 of these having been launched 5 years earlier. A successor, DNSS (Defence Navigation Satellite System), providing continuous position-fixing for all 3 US Forces, is under development. The US Navy has been experimenting with Timation, requiring multiple satellites in high polar orbits of about 9950 × 12,400 miles (16,000 × 20,000 km). The only named launch, Timation 2, on September 30, 1969, placed 6 Timations, together with 2 other satellites, in 563 × 584-mile (906 × 940-km) orbits, with 70° incl. An unnamed launch on December 14, 1971, placed 4 more satellites in similar orbits; Timation 3 was scheduled for late-1973. The US Navy is seeking techniques to give protection against nuclear-weapon radiation and interference by inspector/destroyer satellites. The US Air Force has been working for 7 years on development of groups of 4–5 satellites which would provide aircraft, civil as well as military, with their altitude and speed as well as geographical position. Prototype launches were due in 1973.

NIMBUS

History Originally conceived (as the Latin title-word, meaning 'raincloud' implies) as meteorological satellites to provide atmospheric data for improved weather forecasting; but as increasingly sophisticated sensing devices were added to successive spacecraft, the series grew into a major programme studying earth sciences. From it, too, sprang the ERTS, the results of which proved to be quite sensational, and which are described separately. By mid-1973, data provided by Nimbus 1–5, which was being studied and applied, covered oceanography (the geography of the oceans), hydrology (the study of water in the atmosphere, and on land surfaces, in the soil and underlying rocks), geology (the history of earth, especially as recorded in rocks), geomorphology (the study of the earth, its distribution of land and water and evolution of land forms), geography (description of land, sea and air, and distribution of life, including man and his industries); cartography (chart and map making), and agriculture (data on moisture and vegetation patterns over various land surfaces). By 1973 6 of the 7 planned Nimbus spacecraft had been launched; they were lettered A–F before launch, and given numbers only if successful. All except B, which was lost as a result of a launch-vehicle failure have so far exceeded their mission objectives. F, which should become Nimbus 6, is scheduled for 1974. Total programme costs are estimated at £142 million ($340 million).

Spacecraft Description The basic Nimbus spacecraft is butterfly-shaped when deployed in orbit, 10 ft (3 m) long, with a span across the solar panels of 11 ft (3·4 m). The panels, each 8 × 3 ft (2·4 × 0·9 m) provide more than 200 W of power, supplemented by 2 SNAP–19 nuclear generators. The body of the spacecraft consists of 2 main components, separated by struts; the larger carries the sensors and other equipment; the

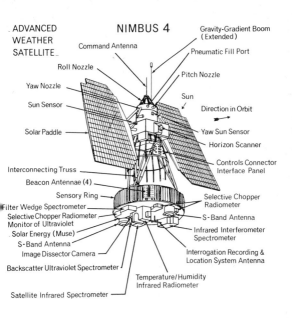

ADVANCED
WEATHER
SATELLITE

NIMBUS 4

Gravity-Gradient Boom
(Extended)

Command Antenna

Pneumatic Fill Port

Roll Nozzle

Pitch Nozzle

Yaw Nozzle

Sun Sensor

Sun

Direction in Orbit

Solar Paddle

Yaw Sun Sensor

Horizon Scanner

Controls Connector
Interface Panel

Interconnecting Truss

Beacon Antennae (4)

Sensory Ring

Selective Chopper
Radiometer

Filter Wedge Spectrometer

S-Band Antenna

Selective Chopper Radiometer
Monitor of Ultraviolet
Solar Energy (Muse)
S-Band Antenna
Image Dissector Camera

Infrared Interferometer
Spectrometer

Interrogation Recording &
Location System Antenna

Backscatter Ultraviolet Spectrometer

Temperature/Humidity
Infrared Radiometer

Satellite Infrared Spectrometer

smaller, between the solar panels, houses the stabilization and control system; working on the principle that the earth is warm and space is cold, the infra-red sensors keep the satellite's cameras pointed towards earth at all times. The system is controlled by a computer which maintains the correct attitude, to within 1 degree in each axis, by means of cold gas jets and inertia wheels. Pitch and roll stability is sensed by horizon scanners, while a gyroscope controls stability in the yaw axis.

Nimbus 1   L. August 28, 1964 by Thor–Agena

B from Vandenberg. Wt 830 lb (376 kg). Orbit 579 × 262 miles (932 × 422 km). Incl. 98°. This was fully operational for 1 month until failure of the solar array power system; a short second burn of the Agena rocket resulted in an eccentric orbit instead of the intended circular 690-mile (1110-km) orbit. It returned 27,000 cloud-cover photos, providing the first high-resolution TV and infra-red weather photos, and proving to meteorologists that such pictures could be received at small, inexpensive portable stations as the satellite passed overhead. Hurricane Cleo's 'portrait' was taken during Nimbus 1's first day in orbit; subsequently many other hurricanes and Pacific typhoons were tracked; inaccuracies on relief maps were corrected, and the Antarctic ice front more accurately defined as a result of its pictures. Orbital life 15 years.

Nimbus 2 L. May 15, 1966, by TAT–Agena B from Vandenberg. Wt 912 lb (413 kg). Orbit 684 × 734 miles (1100 × 1181 km). Incl. 100°. A nearly perfect orbit, designed for 6 months (2500 orbits), actually provided 33 months of operations and terminated in orbit 13,029 on January 17, 1969. Following the success of Nimbus 1, over 300 APT (Automatic Picture Transmission) stations in 43 countries were able to receive its high-resolution infra-red pictures. Temperature patterns were obtained of lakes and ocean currents for shipping and fishing industries, and thermal pollution could be identified. An additional Medium Resolution Infra-red Radiometer, which measured electromagnetic radiation emitted and reflected from earth in 5 wavelength intervals from visible to infra-red, permitted detailed study of the effect of water vapour, carbon dioxide and ozone on the earth's heat balance. Orbital life 800 years.

Europe as seen from Nimbus 3 on orbit 1530, Aug 6, 1969

Nimbus 3   L. April 14, 1969, by Thorad–Agena
D from Vandenberg. Wt 1269 lb (575 kg). Orbit
665 × 703 miles (1070 × 1131 km). Incl. 99°. The
third success in the series following the launch
failure of Nimbus B on May 18, 1968. Nimbus 3
carried 7 meteorological experiments, plus SNAP–

19, a nuclear-power unit for generating electricity. Vertical temperature measurements of the atmosphere by SIRS (Satellite Infra-red Spectrometer) were acclaimed as one of the most significant developments in the history of meteorology. Previously, because data over the oceans had been scanty, only 20% of the world had detailed weather information; SIRS made it possible to obtain temperature data over the entire earth with an accuracy of 2°F above 20,000 ft (6100 m), and within 4°F below that level. A 23-lb (10·4-kg) electronic collar fitted to a wild elk in the National Elk Refuge in Wyoming, was interrogated twice daily to study the migratory habits of large animals. The Nimbus 3 rocket also placed Secor 13, a US army geodetic satellite, weighing 45 lb (20·4 kg) into an identical orbit. Life of Nimbus 3 is 800 years; Secor 2000 years.

Nimbus 4   L. April 8, 1970 by Thorad–Agena D from Vandenberg. Wt 1488 lb (675 kg). Orbit 679 × 688 miles (1093 × 1107 km). Incl. 107°. In its circular, near-polar orbit, Nimbus 4 was still making world-wide weather observations on a twice daily basis (once in daylight and once in darkness) more than 2 years later. Of its 9 experiments, 4 were new and 5 improved versions of experiments on earlier flights. New experiments included IRLS (Interrogation Recording and Location System); examples of its tracking activities included weather balloons floating around the world, floating ocean buoys, a wild animal, and Miss Sheila Scott on a solo round-the-world flight. One task was to measure the thickness of ice on floating islands in the Arctic by interrogating buoys placed in the water; one such island, T3, was 7 × 4 miles (11 × 6 km) long and 100 ft (30 m) thick. These islands melt in summer in brackish swamps covered by fog and are impossible to reach. Details of their behaviour are providing information on the 'cradle' of much of the weather affecting the US and Europe. Other Nimbus 4 tasks included the analysis of water qualities and sewage pollution of the Great Miami River near Cincinnati and of Lakes Erie and Ontario. Low oxygen content indicates raw effluent, the oxygen decrease being caused by bacterial decomposition. Other IRLS 'platforms' were also placed in Mt Kilauea, Hawaii, probably the world's most active volcano, to measure the relationship between temperature rises and eruptions, and in a bear's den in Montana to monitor its environment during hibernation. Each 'platform' will only respond when the satellite interrogates it by its individual 16-bit digital 'telephone number'; the address of the bear den in Montana is 0111100001111001. The Nimbus 4 rocket also placed TOPO 1, a US Army Secor-type satellite, wt 48 lb (21·7 kg), into an identical orbit for use in space-ground tactical exercises. Orbital life: Nimbus 4 1200 years; Topo 1 2000 years.

Nimbus 5 L. December 11, 1972 by Delta from Vandenberg. Wt 1695 lb (768 kg). Orbit 677 × 684 miles (1089 × 1102 km). Incl. 99°. The heaviest in the series, this carried 6 advanced instruments or sensors—3 new ones, and 3 improved versions of earlier equipment. Nimbus 5 was designed to take the first vertical temperature and water readings of earth's atmosphere through cloud—a major step forward, since many parts of the world are under cloud-cover for more than 50% of the time. Sensors were also intended to map the Gulf Stream off the US East Coast and the Humboldt Current off South America's West Coast. Plotting the daily position of the Gulf

Stream enables south-bound ships to avoid it, while northbound ships, by riding in the stream, get the benefit of several extra knots. Operational life 1 year. Orbital life 1600 years.

ORBITING ASTRONOMICAL OBSERVATORIES—OAO

History A programme of 4 Orbiting Astronomical Observatories, begun in 1959, and completed, with the launch of the fourth (on August 21, 1972) at a total cost of £154·6 million ($364·4 million). They were intended to seek answers to fundamental questions concerning interstellar matter, as well as the stars themselves. They culminated in the heaviest and most complex spacecraft launched by NASA to that date; once in orbit it was renamed Copernicus, as part of the 500th anniversary in 1973 of the birth of the Polish astronomer, widely considered 'father of modern astronomy'.

OAO 1 L. April 8, 1966 by Atlas–Agena D from Cape Kennedy; wt 3917 lb (1776 kg). Orbit 492 × 500 miles (792 × 805 km). Incl. 35°. The battery failed after only 3 days in orbit, but it provided engineering data providing that the concept of astronomical investigation from space was feasible, and led to improvements in later craft.

OAO 2 L. December 7, 1968, by Atlas–Centaur from Cape Kennedy; wt 4446 lb (2016 kg). Orbit 479 × 485 miles (770 × 780 km). Incl. 34·99°. At the launch of OAO–C, it was still operating, far beyond its expected lifetime; its 11 telescopes had made more than 10,000 observations of 1500 celestial objects, and provided astronomers for

the first time with the opportunity to conduct sustained viewing of the universe. Its ultra-violet instruments gave new insights into stars, galaxies, and solar system, and the earth's upper atmosphere; they made some major observations of comets, and, in May 1972, of a supernova—the momentary outburst of a star to a brightness millions of times greater than the sun.

OAO-B L. November 30, 1970 by Atlas–Centaur from Cape Kennedy. It failed to achieve orbit when a protective shroud could not be jettisoned, and fell back to earth.

OAO-C L. August 21, 1972, by Atlas–Centaur from Cape Kennedy; wt 4900 lb (2220 kg). Orbit 465 × 460 miles (748 × 740 km). Incl. 35°. The heaviest scientific payload ever launched by the US. At the centre of the 10-ft (3-m) craft is the main 32-in. (80-cm) ultra-violet telescope; designed by Princeton University, its mirror is made from thin fused silica ribs, and weighs only 105 lb (47·6 kg), compared with 360 lb (163 kg) of a conventional glass mirror. It is capable of observing an object the size of a football at a distance of 400 miles (650 km). Ultra-violet light from a star is collected and directed to a spectrometer which then sends digital readings to earth. Observations can be made of clouds of interstellar gas. 3 smaller X-ray telescopes, developed at University College, London, and sponsored by the Science Research Council, will be looking for sources of X-rays other than the sun; earlier satellites, notably Explorer 42, L. December 1970, have established that X-rays are present in the universe in much larger amounts than are generated by the sun. It is hoped the British experiment will chart about 200 other X-ray sources already identified, and find new sources. The spacecraft is also examining the theory that some objects obtain their heat from gravity rather than nuclear fusion. Copernicus, which is expected to last about a year, cost about £32.6 million ($78.2 million).

PIONEER

Pioneer 10

Lift-off Wt Overall: 323,415 lb (146,673 kg)
Lift-off Ht: 132 ft (40·3 m)
Spacecraft Wt Overall: 570 lb (260 kg)
Spacecraft Wt Sc. Payload: 61·5 lb (27·8 kg)
Spacecraft Radius: 21 ft (6·4 m)
Spacecraft Antenna Dia.: 9 ft (2·7 m)
Launch Vehicle: Atlas–Centaur–TE–M–364–4

History   With its latest craft being used to open up the solar system for future manned flights, Pioneer has become the most exciting of all the unmanned projects. Pioneer 1, launched in 1958, was the first NASA spacecraft, the project having been handed over by the USAF. The series is complementary to the Mariner and Viking flights. Missions are given letters before launch, changed to the next number in the series only if successfully placed on course. In 15 years, 11 out of 14 launches added immensely to man's knowledge of the earth's surroundings, and most were still transmitting. The first 5 were used to study solar energy, and also to provide up to 15 days' warning of solar flares, so that astronauts could be protected during moonflights. Pioneers 6–9, still operating at the time of writing, were used to study space from widely separated points during an entire solar cycle of 11 years; the first of these, designed for a 6-month lifetime, had been operating for nearly 8 years. A major achievement of this series was to discover the earth's long magnetic 'tail', about 3·5 million miles (5·6 million km), on the side away from the sun. In 1971 the alignment of Pioneers 6 and 8, at points more than 100 million miles (161 million km) apart, enabled the solar wind's density to be measured more accurately than ever before. In September 1972 NASA's Ames Research Centre in California, succeeded in locating and reviving Pioneer 7, which had turned its transmitters off, although it was on the far side of the sun, 194 million miles (312 million km) from earth.

Spacecraft Description   Pioneer spacecraft are designed to carry a variety of instruments tailored to the study of either individual or a series of planets. Nos. 10 and 11 were the first craft designed to travel into the outer solar system, to operate there for 7 years, and as far from the sun as 1500 million miles (2400 million km). Since launch energy requirements to reach such distances are far higher than for shorter missions, the spacecraft must be very light; the 570 lb (258 kg) weight includes 65 lb (30 kg) of scientific instruments, and 60 lb (27 kg) of propellant for attitude changes and midcourse corrections. 6 thrusters each provide 0·4–1·4 lb (0·2–0·6 kg) thrust. They can adjust the place and time of arrival at Jupiter by changing velocity, or merely adjust the attitude. This is done by pulse thrusts, timed by a signal from a star sensor which 'sees' Canopus once per rotation, or by one of the 2 sun sensors which 'see' the sun once per rotation. Velocity changes totalling 420 mph (670 kph) can be made during the mission. The spacecraft are spin-stabilized, giving the instruments a full circle scan 5 times a minute. Because solar radiation at Jupiter is too weak to operate an efficient solar-powered system, 4 nuclear units provide electric power; they are carried on 9-ft (2·7-m) booms so that their radiation will not affect the scientific experiments (13 on Pioneer 10; 14 on Pioneer 11). Controllers use 222 different commands to operate the spacecraft; during the 4 days it takes to pass Jupiter commands take 45 mins to reach it. The heart of the communications system is the fixed, 9-ft (2·7-m) dia. dish antenna, which focuses the radio signals in a narrow beam. The onboard experiments are intended to return data on the solar atmosphere from earth to beyond Jupiter; to measure the numbers and characteristics of asteroids; to measure Jupiter's atmosphere, heat balance, and internal energy sources; its magnetic fields and radiation belts; and, by means of the imaging photo polarimeter, to return images

(*left*) Atlas/Centaur starts Mariner 7 on its 4-month flight to Mars on Mar 27, 1969

Plate 1

(*right*) Night launch by Titan 3C, on Nov 2, 1971, of America's top-secret 'Big Bird' spy satellite

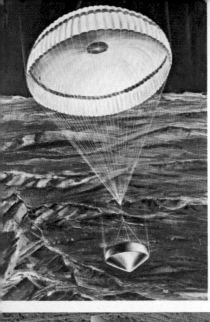

(*left*) Artist's conception of US Viking descending on Mars in mid-1976

Plate 2

(*below*) Viking as it will sample Martian soil; sensor on upper right measures wind and temperature

(*above*) Pioneer 10 flew within 81,000 miles
(130,300 km) of Jupiter on Dec 4, 1973
Plate 3 (Artist's conception)

(*below*) Pioneer 10: final assembly

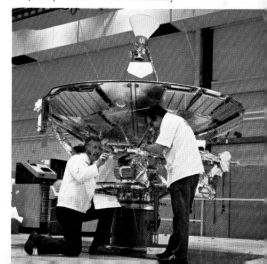

(*left*) Ariel 4, preparing for Spin Balance Tests

Plate 4

(*below*) Comsat Control Centre, Washington, D.C. (*see* INTELSAT)

(*left*) Europa 2, just before the final failure in Nov 1971

Plate 5

(*below*) ERTS-1 has sent back over 100,000 pictures

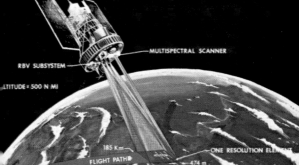

MULTISPECTRAL SCANNER

RBV SUBSYSTEM

LTITUDE = 500 N MI

185 Km

HOUSTON

ONE RESOLUTION ELEMENT

FLIGHT PATH

474 m

Plate 6 ERTS picture showing New York in upper right corner: Philadelphia, lower centre, on Delaware River. 10% of US population lives in this area

Plate 7 ERTS picture of Grand Teton National Park and Idaho area. Teton Range, upper centre; Snake River Plain, upper left. Clear water appears black. Both views from 568 miles (906 km) altitude

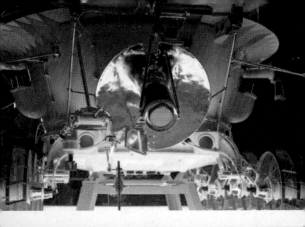

Plate 8 Lunokhod 1 at Paris Air Show: (*above*) moon's-eye view of panoramic TV camera at rear; (*below*) front view, showing TV 'eyes'; narrow-beam antenna, top right; omni-directional antenna (cone-shape) left

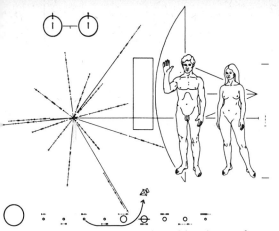

Bottom, symbols of sun and planets show origin and course of Pioneer 10. *Top left*, hydrogen atom, used as universal clock in conjunction with pulsar lines and frequencies, indicates when and where spacecraft was launched. Size of figures shown against spacecraft outline

of Jupiter. Pioneers 10 and 11 both carry a 6 × 9-in. (15 × 23-cm) plaque showing the origin of the spacecraft in the solar system, and drawings of a man and woman related to the spacecraft's size in case they should one day be seen by another intelligent species; but such a species would need to have eyes like our own to be able to understand the plaque's message of goodwill.

Pioneer 1   L. October 11, 1958 by Thor–Able from Cape Canaveral. Wt 84 lb (38 kg). Failed to reach the moon, but looped 70,717 miles (113,854 km) into space. In 43 hrs 17 mins of flight, it discovered the extent of the earth's radiation bands.

Pioneer 2   L. November 8, 1958, by Thor–Able

from Cape Canaveral. This 87-lb (39·5-kg) probe, intended to reach the moon, failed as a result of unsuccessful third-stage ignition.

Pioneer 3 L. December 6, 1958, by Juno 2 from Cape Canaveral. The 13-lb (5·9-kg) probe again failed to reach the moon, but reached height of 63,580 miles (102,333 km), and discovered earth's second radiation belt.

Pioneer 4 L. March 3, 1959, by Juno 2 from Cape Canaveral. 13-lb (5·9-kg) lunar probe, passed within 37,300 miles (59,983 km) of moon, and then into solar orbit 0·9871 × 1·142 AU.

Pioneer 5 L. March 11, 1960 by Thor–Able from Cape Canaveral. 95-lb (43-kg) probe, sent into solar orbit 0·8061 × 0·995 AU, and sent back solar flare and wind data until June 26, 1960, at a distance of 22·5 million miles (37 million km).

Pioneer 6 L. December 16, 1965 by TAD from Cape Kennedy. 140-lb (63-kg) cylinder, 37 in. (1 m) dia. and 35 in. (0·9 m) high, launched towards sun into 0·814 × 0·985 AU solar orbit. It sent back first detailed description of the tenuous solar atmosphere; with Pioneer 7, gathered continuous data on events on a strip of solar surface extending nearly halfway round the sun.

Pioneer 7 L. August 17, 1966, by TAD from Cape Kennedy. 140 lb (63 kg) cylinder, launched away from sun into 1·010 × 1·125 AU solar orbit. (*See* Pioneer 6).

Pioneer 8 L. December 13, 1967, by TAD from Cape Kennedy. 145 lb (65·3 kg) cylinder, placed

in 1·0 × 1·1 AU solar orbit, to join previous 2 Pioneers in obtaining data on solar wind, magnetic field and cosmic rays. Also defined tail of earth's magnetosphere. Launched piggy-back with Pioneer 8, was NASA's first Test and Training Satellite, used to exercise the Apollo communications network.

Pioneer 9 L. November 8, 1968 with 2nd Test and Training Satellite, by TAD from Cape Kennedy. 148-lb (66·6-kg) Pioneer placed in 0·75 × 1·0 AU solar orbit. This was 4th of what was to be a series of 5 solar probes; but Pioneer E, intended to be Pioneer 10, carrying 3rd piggy-back Test Satellite, failed to orbit on August 27, 1969.

Pioneer 10 L. March 3, 1972, by Atlas-Centaur from Cape Kennedy. Wt 570 lb (270 kg). Initial speed 32,000 mph (51,800 kph), faster than any previous man-made object, placed it on a trajectory which took it past Jupiter's cloud-tops at a distance of 81,000 miles (130,300 km) on December 4, 1973. Initial speed was achieved by adding for the first time to an Atlas-Centaur booster (lift-off thrust 411,353 lb; 186,590 kg), a solid-fuelled 3rd-stage TE-M-364-4, developing 15,000 lb (6800 kg) thrust. (This is an uprated version of the retromotor used for the Surveyor moon lander.) 4 days after launch, 2 brief firings of the spacecraft's thrusters (8 mins 7 secs and 4 mins 16 secs) increased velocity by 45.9 ft (14 m) per sec, shortened the flight time to Jupiter by 9 hours, and adjusted the arrival point to 14° below the Jovian equator. From November 26, 1973, by which time command-and-return radio signals took 90 mins, a total of about 300 pictures were obtained of the approach and fly-by. Early ones were mainly for calibration, but about 40 are expected to yield much information after lengthy computer processing. First results (this is a last-minute entry in *Unmanned Spaceflight*) suggested that the darker coloured bands around Jupiter are lower and warmer, with the white ones being higher and colder belts of cloud. The primary objectives of defining the magnetic field and radiation belts were achieved, and much discovered about its atmosphere and mass. Good pictures were obtained of the Great Red Spot, but early analysis provided neither confirmation nor disproof of the theory that it is a column of gas. The biggest disappointment was failure to obtain good pictures of the orange moon, Io. Nevertheless, other data confirmed that it had a single-layer atmosphere (earth's is multi-layered), is probably composed of rock and iron, and is probably more earthlike than Jupiter itself. Some pictures were obtained of Ganymede, Callisto and Europa;

Ganymede was already known to have an atmosphere; the other two are now also believed to have tenuous atmospheres.

Pioneer 10 was the *first* spacecraft placed on a trajectory to escape from the solar system into interstellar space; the *first* to fly beyond Mars; the *first* to enter the Asteroid Belt; and the *first* to fly to Jupiter. It is hoped it will be the first to sense the interstellar gas beyond the sun's atmosphere. It should reach its absolute communications limit, near the orbit of Uranus in 7½ years after launch, at about 1800 million miles (2900 million km) from the sun; it is expected to leave the solar system in 1987.

To achieve all this, however, Pioneer 10 needed to survive 2 major hazards. First, the possibility of collision during its passage through the 175-million-mile (280-million-km) thick Asteroid Belt in July, 1972. This is a band of cosmic rubble circling the sun between the orbits of Mars and Jupiter. The largest asteroids reach nearly 500 miles (805 km) in dia., ranking as minor planets; orbits have been calculated for 1776 asteroids, and there may be 50,000 ranging from the largest, Ceres, 480 miles (770 km) dia. down to bodies of 1 mile (1·6 km) dia., and uncountable fragments below that. Scientists consider the Belt is debris from the breakup of a small planet; total volume of the material would make a planet about one-thousandth the size of earth. In the event, Pioneer 10 emerged safely from the asteroid belt on February 15, 1973. Preliminary findings suggested that the belt contained much less material than previously thought, and presented little hazard to spacecraft.

Pioneer 10 also survived the second hazard—the Jovian radiation belts, far more intense than earth's. Some of its instruments however, particularly its asteroid-meteoroid detector, were degraded. Though scarred, Pioneer 10 survived so well that it was decided that a closer approach by Pioneer 11 would be possible.

Pioneer 11 L. April 5, 1973, by Atlas–Centaur from Cape Kennedy. Identical to Pioneer 10, apart from the addition of a 2nd magnetometer to measure Jupiter's high magnetic fields. An on-time launch on its 609-day, 630-million-mile (1013-million-km) journey, means that for a year after its launch final decisions on its trajectory and mission can be kept open, depending upon the success of Pioneer 10. If it had failed, Pioneer 11 would have repeated the mission; alternatively its trajectory can be altered to enable it to fly a different course over Jupiter's surface, possibly approaching as close as 27,000 miles (43,500 km) above the striped cloud-tops; investigating a second of the 12 moons; and then, perhaps swinging on to Saturn in 1980. Earliest arrival date at Jupiter is December 5, 1974.

Future Missions 4 Pioneer missions to explore Venus are being planned, starting in January 1977. The first 2 missions would be launched within a few days, and each would carry 4 probes. These would detach and investigate the upper atmosphere. A Venus orbiter is planned for 1978, and another probe mission for 1980. Venus is earth's closest neighbour in the solar system, similar in size, and possibly in origin. Its low rotation rate, apparently complete cloud cover, extremely dense atmosphere, and high surface temperature, make Venus the object of intense scientific interest. A detailed comparison of its nature and composition with that of the earth's would lead, it is thought, to better predictions of atmosphere changes on earth.

Jupiter

Much the largest of the 9 planets, with an equatorial dia. of 88,900 miles (143,000 km)—earth's is 8000 miles (12,875 km). Its orbit lies 480 million miles (773 million km), from the sun, which it circles once in just under 12 years, though it rotates once in 10 hrs, the shortest day of any of the planets. Because of its size, this means that a point on the equator races along at 22,000 mph (35,400 kph) compared with a speed of 1000 mph (1600 kph) for a similar point on earth. This rotational speed, combined with Jupiter's fluid character, results in a bulge at the equator; the polar dia. is only 77,000 miles (124,000 km), or 11,800 miles (19,000 km) smaller than the equator. The nature of the Jovian surface is quite unknown, because so far it has not been possible to penetrate the thick atmosphere. Most scientists say the planet is made up of a mixture of elements similar to that in the sun, or the primordial gas cloud which formed the sun and the planets. At least three-quarters is hydrogen and helium. The view through a telescope is almost certainly of the tops of towering, multi-coloured clouds. The planet is banded parallel with the equator, with grey, salmon-coloured and slate-grey stripes which change hue periodically, possibly as a result of the sun's 11-year activity cycles. The Great Red Spot in the southern hemisphere seems to brighten and darken at 30-year intervals. The cloud-layers are variously estimated at between 60–3600 miles (100–6000 km) deep. Such a deep, dense atmosphere may mean a surface pressure 200,000 times that of earth's. The atmosphere is known to contain ammonia, methane and hydrogen, which, along with water, are described by NASA as 'the chemical ingredients of the primordial "soup" believed to have produced the first life on earth by chemical evolution'. On this evidence Jupiter could contain 'the building blocks of life'. This would probably consist of low-energy life forms, such as plants and micro-organisms, because there is believed to be no free oxygen. The life forms would probably float or swim in the atmosphere, because of the very high pressures at the surface.

Jupiter is the only planet apart from earth known to have a magnetic field; and earth receives more radio noise from Jupiter than from any source except the sun. Regular decametric bursts of radio noise, believed to originate in huge discharges of electricity, are equal to the power of several hydrogen bombs. And though only about $\frac{1}{27}$th as much heat from the sun arrives at Jupiter as arrives at earth, the planet radiates $2\frac{1}{2}$–3 times more energy than it absorbs. Jupiter has 12 moons, 2 of which, Ganymede and Callisto, are about the size of the planet Mercury, 2 others, Io and Europa, are similar in size to the earth's moon; 7 tiny outer moons have diameters ranging from 80 miles (130 km), down to 9 miles (15 km). At least one moon, Io, has an atmosphere.

RANGER

Rangers 6–9
$\begin{cases} \text{Wt 806 lb (366 kg)} \\ \text{Ht 10 ft 3 in (3·1 m)} \\ \text{Solar Panels: 15 ft (4·5 m)} \\ \text{Launcher: Atlas–Agena D} \end{cases}$

History   The first of 3 projects (the others being Lunar Orbiter and Surveyor) aimed at obtaining sufficient knowledge and pictures of the lunar surface to make the manned Apollo landings possible. The first 6 missions were failures,

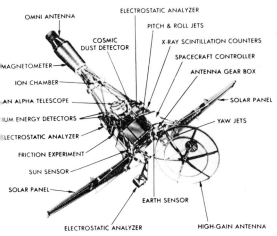

OMNI ANTENNA

MAGNETOMETER

ION CHAMBER

AN ALPHA TELESCOPE

IUM ENERGY DETECTORS

ELECTROSTATIC ANALYZER

FRICTION EXPERIMENT

SUN SENSOR

SOLAR PANEL

ELECTROSTATIC ANALYZER

COSMIC DUST DETECTOR

ELECTROSTATIC ANALYZER

PITCH & ROLL JETS

X-RAY SCINTILLATION COUNTERS

SPACECRAFT CONTROLLER

ANTENNA GEAR BOX

SOLAR PANEL

YAW JETS

EARTH SENSOR

HIGH-GAIN ANTENNA

Ranger 1

beginning with Ranger 1, launched from Cape Kennedy on August 23, 1961, right through to Ranger 6, launched on January 30, 1964. Plans to eject and hard-land a 94-lb (42·6-kg) capsule to measure seismic activity on the moon's surface as a mission 'bonus' were abandoned, as failures continued right through the manned Project Mercury flights, and the need for close-up pictures of the lunar surface became desperate if Project Apollo was not to be held up. For Rangers 6–9 a 375-lb (170-kg) conical structure containing 2 wide-angle, and 4 narrow-angle TV cameras, plus video combiner, transmitters, etc, was mounted on the spacecraft to send back about 14 mins of TV pictures from 1405 miles (2261 km) above the moon, until the spacecraft was destroyed by impact at 5800 mph (9330 kph). Ranger 6 looked like being successful, until it was found that the

TV system had been destroyed by being inadvertently turned on early in the flight. Rangers 7, 8 and 9, however, justified the persistence of NASA's Jet Propulsion Laboratory at Pasadena, California. Ranger 7, launched on July 28, 1964, after a 68-hr flight, returned 4308 excellent pictures—man's first close-up lunar views—before impacting in the Sea of Clouds. Ranger 8, launched on February 17, 1965, after a 64-hr flight, returned 7137 high-quality photographs; they covered 900,000 square miles (2,331,000 sq km) in the Sea of Tranquillity area. Finally Ranger 9, launched on March 21, 1965 after a 64-hr flight covering 259,143 miles (417,054 km), relayed 5814 excellent pictures before impacting within 3 miles (4·8 km) of the target point in Crater Alphonsus. Photographs covered features down to only 10 in. (25 cm) across; 200 of the pictures were shown 'live' on TV—probably the first live 'TV spectacular' from the moon. The total of over 17,000 pictures, some revealing craters as small as 10 in. (25 cm) across, revealed the lunarscape several thousand times better than the best earth telescope had ever done, and finally justified Ranger's cost of £108 million ($260 million). But while they provided the evidence needed to complete the design of soft-landing spacecraft, it remained for Project Surveyor to establish whether the moon's surface was firm enough to support manned landings.

Spacecraft Description   Conical camera structure, mounted on hexagonal spacecraft bus. Battery of 6 TV cameras (2 wide-angle, 4 narrow-angle) covered by polished aluminium shroud, with 13-in. (33-cm) opening near top. High-gain antenna, and 2 solar panels hinged to base. Midcourse motor provided 50-lb (22·6-kg) thrust for up to 98 secs. Attitude control, maintained by nitrogen gas jets, sun and earth sensors and 3 gyros, programmed by computer and sequencer, enabled spacecraft to aim cameras at moon while pointing high-gain antenna to earth for transmission.

SURVEYOR

History The third of the 3 unmanned lunar exploration projects carried out in parallel with the 3 manned projects aimed at placing men on the moon before 1970. Surveyors 1, 3, 5 and 6, successfully soft-landed at sites spaced across the lunar equator, achieved all Apollo objectives, enabling Surveyor 7 to be landed on the rim of Crater Tycho, to conduct digging, trenching, and bearing tests etc.

Surveyor 1, launched on June 1, 1966, from Cape Kennedy, after a 63-hr 36-min flight, made the world's first fully controlled soft-landing in the Ocean of Storms; in the following 6 weeks, it sent back 11,150 pictures, from horizon views of mountains to close-ups of its own mirrors, etc.

Surveyor 2, launched on September 20, 1966, crashed S.E. of Crater Copernicus when one of the vernier engines failed to fire.

Surveyor 3, launched April 17, 1967, landed safely despite a heavy bounce, in the Ocean of Storms 380 miles (612 km) E. of Surveyor 1. In addition to returning 6315 photos, it used a scoop to make the first excavation and bearing test on an extraterrestrial body.

(In November 1969 Apollo 12 landed almost alongside Surveyor 3, and Conrad and Bean brought back to earth the TV camera, scoop, and other parts, for studies to be made of the effects of 31 months of lunar exposure.)

Surveyor 4, launched July 14, 1967, lost radio contact $2\frac{1}{2}$ mins before touchdown, and crashed in Sinus Medii (Central Bay).

Surveyor 5 was launched September 8, 1967, and major technical problems were successfully solved by tests and manoeuvres during the flight; 18,000 photos were obtained from the southern part of the Sea of Tranquillity, and the first on-site chemical soil analysis was carried out.

Surveyor 6, launched November 7, 1967, landed in Sinus Medii in the centre of the moon's front face; in addition to sending over 30,000 pictures, it performed the first take-off from the lunar surface; its 3 vernier engines, fired for $2\frac{1}{2}$ secs, lifted it to 10 ft (3 m), and landed it again 8 ft (2·4 m) away—a test providing confidence that the surface was firm enough for a manned landing.

Surveyor 7, launched January 7, 1968, was then sent to Crater Tycho; chemical analyses suggested

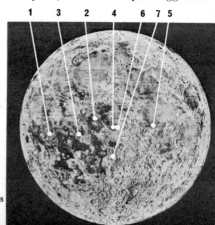

Surveyor landing points

that the debris there had once been in a molten state. 21,000 photos were obtained; but Apollo astronauts' requests to visit this area were turned down because of the extra fuel needed to reach and return from such a remote area.

Spacecraft Description A 3-legged vehicle 10 ft (3·05 m) high, with a triangular aluminium frame, providing mounting surfaces and attachment points for landing gear, main retro-rocket engine of up to 10,000 lb (4536 kg) thrust, and 3 vernier engines of up to 104 lb (47 kg) thrust, etc. Launch wt of 2200 lb (998 kg) was reduced by fuel consumption to 625 lb (283 kg) on landing. A central mast supported the high-gain antenna and single solar panel of 30 × 44 in. (0·75 × 1 m). Aluminium honeycomb footpads were attached to each leg of the tripod landing gear. The TV camera was pointed at a mirror which could swivel 360°, and at earth command be focused from 4 ft (1·2 m) to infinity with narrow or wide-angle views, returning 200 and 600-line photos. Surveyors did not go into lunar orbit before landing; in a direct approach, automatically controlled by radar, the main retro-rocket was fired at an altitude of about 60 miles (96 km) for 40 secs. At 25 miles (40 km) altitude, with speed down to 250 mph (402 kph), the verniers took over, dropping the craft on to the surface, in the case of Surveyor 1, from 14 ft (4·3 km) at 7·5 mph (12 kph). Launcher: Atlas–Centaur.

TIROS/ESSA

History TIROS (Television and Infra-Red Observation Satellite) began as a joint NASA/Defense Department project to develop a meteorological satellite. As soon as Tiros 1, L. April 1, 1960, began orbiting at 430 × 460 miles (692 × 740 km), it was clear that the US had successfully established both a meteorological survey and military reconnaissance satellite. During the 78 days that its batteries lasted, it sent back 22,952 cloud-cover photographs; as the first of them came in, a meteorologist at a ground station observed that the programme had gone 'from rags to riches overnight'. It is believed that its photographs included some of the Soviet Union and China, so detailed that aircraft runways and missile sites could be readily identified. By the time 10 had been launched, the more advanced Nimbus and ESSA satellites were taking over; the first 8, all operating in similar orbits, sent back several hundred thousand photographs, together with information about the flow of heat the earth was reflecting back into space—vital meteorological information unobtainable until then. Tiros 9, L. January 22, 1965, was the first attempt to reach polar orbit from Cape Kennedy; the series of 3 Delta 'dog-leg manoeuvres' duly placed it in the planned 82° sun-synchronous orbit; but due to a 2nd-stage failure to cut-off, the orbit, instead of being 400 miles (644 km) circular, was 435 × 1602 (700 × 2578 km). In the event, the higher apogee provided more earth cover than planned; on February 13, the first 'photomosaic' of the entire world's cloud cover was provided by 450 excellent pictures. By the time Tiros 10, L. July 2, 1965, was shut down on July 3, 1967, more than 500,000 cloud-cover pictures had been returned, and the system had become the basis of the first operational weather satellite system called TOS (Tiros Operational Satellite) and managed by ESSA (Environmental Science Services Administration) which, between Feb-

ruary 3, 1966, and February 26, 1969, launched a series of 9. In 1969 a picture from Essa 7 made history by revealing that the snow cover over America's mid-west, in Minnesota and the Dakotas, was 3 times thicker than normal. Measurements showed that it was equivalent to 6–10 in. (15–25 cm) of water covering thousands of square miles. A disaster area was declared before it happened; and when the floods came much had been done to control the situation. By then, 305 stations in 90 countries were receiving the weather pictures, as well as an unknown number of private citizens who had built their own receivers.

ITOS/NOAA 1970 saw the start of an **I**mproved **T**iros **O**perational **S**atellite System; ITOS satellites were given a new series of NOAA designations once in orbit, because ESSA had been taken over by the National Oceanic and Atmospheric Administration. NOAA 1, L. December 11, 1970, by Delta from Vandenberg, in an 888 × 915-mile (1429 × 1472-km), orbit with 101° incl., worked well; but NOAA 2, L. October 21, 1971, failed to achieve orbit.

Spacecraft Description Tiros satellites have a 'hatbox' shape; 18-sided polygons, 42 in. (1·07 m) dia., and 22 in. (0·55 m) high. Solar cells cover the sides and top, with apertures for 2 TV cameras on opposite sides; each camera can take 16 pictures per orbit at 128-sec. intervals, though the interval can be decreased to 32 secs. 2 tape recorders can store up to 48 pictures when ground stations are out of range. The weight of 263 lb (119 kg) for Tiros 1 had risen to 305 lb (138 kg) for Tiros 9 and 10. Essa satellites, similar but more advanced, have 2 APT cameras able to photograph a 2000-mile (3218-km) wide area, with 2-mile (3·2-km) resolution at picture centre. Pictures are taken and transmitted every 352 secs, allowing a typical APT station to receive 8–10 per day.

VANGUARD

History The Vanguard rocket, a US Navy development of sounding rocket technology built on Viking and Aerobee, was selected in 1955 as the most suitable launcher for America's first satellite —an unfortunate choice, as it turned out, since Russia had launched Sputnik 1 and the US Army had launched Explorer 1 before the first Vanguard satellite went into orbit. The choice had been whether to adapt sounding rockets or military rockets—in this case the US Army Redstone—for

Vanguard 3
exploding, Dec
6, 1957

the first satellite launch. Russia chose military
rockets in the same year, and got there first; in
America, against a background of bitter Service
rivalry, with the US Air Force advocating the use
of their Atlas rocket, the deciding factor was that
development of military rockets, known to be
lagging behind Russia's, must not be held up.

The 1st stage of the Vanguard's rocket, using
liquid oxygen and kerosene, developed 27,000 lb
(12,250 kg) thrust. The 2nd stage, burning white
fuming nitric acid and unsymmetrical dimethyl
hydrazine, provided 7500 lb (3402 kg) thrust.
The 3rd stage, with solid propellant, added up to
3100 lb (1406 kg). It was 72 ft (22 m) long, with
45 in. (1·1 m) dia., and weighed 22,600 lb
(10,250 kg). The first full-scale Vanguard Test

113

Vehicle, designated TV2, with dummy upper stages, was successfully launched from Cape Canaveral on October 23, 1957, sending a 4000-lb (1814-kg) payload on a 109-mile (175-km) high 335-mile (491-km) trajectory. But TV3 toppled over on the launch-pad and exploded. 2 months later, the backup vehicle veered off course and broke up at an altitude of 4 miles (5·8 km). By then Sputnik 1 was in orbit. TV4 finally became Vanguard 1, and a remarkably long-lived and successful satellite, transmitting temperatures and geodetic measurements until March 1964. But TV5 failed; so did the first 3 operational Vanguard Satellite Launch Vehicles, designated SLV1, 2 and 3. There were 2 more failures between Vanguards 2 and 3 and the programme was over. Dr Wernher von Braun, rejoicing in the triumph of Redstone, later said Vanguard was a 'goat' through no fault of its own. For all its failures, Vanguard's 2nd and 3rd stages were later bequeathed to Thor and Atlas, and its 3rd stage also to Scout; and the swivelling motors on its 1st stage worked perfectly on Saturn.

VIKING

History Originally scheduled for 1973, 2 Viking spacecraft are now to be launched during the 30-day Mars 'window' between mid-August and mid-September, 1975. If things go well, the first will make a soft-landing on the 200th anniversary of the United States' achieving independence, on July 4, 1976, to start an ambitious plan to photograph the surface and try to establish whether any form of life exists on the Red Planet. The project continues the Martian exploration already carried out by Project Mariner; 2 landing sites have been chosen as a result of Mariner 9's mapping of Mars in 1972. The first is in a valley known as Chryse (19·5°N, 34°W) at the north-east end of the giant, 3000-mile (4800-km) long Martian Grand Canyon discovered by Mariner 9. About 16,000 ft (5 km) lower than the mean surface, this area may once have been a drainage basin for a large portion of equatorial Mars. The second site is Cydonia, about 1000 miles (1600 km) NE of the first site, in the Mare Acidalium region (44·3°N, 10°W), at the edge of the southernmost reaches of the north pole 'hood'—a hazy veil which shrouds each polar region during the winter season. The hoods may carry moisture, thus increasing the chances of finding evidence of life; and this area is even lower than Chryse—18,000 ft (5486 km) below the mean surface. Both sites are fairly smooth, calm areas.

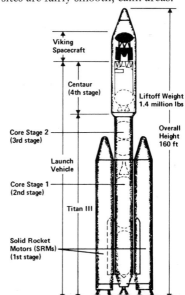

Viking Spacecraft

Centaur (4th stage)

Liftoff Weight 1.4 million lbs

Core Stage 2 (3rd stage)

Overall Height 160 ft

Launch Vehicle

Core Stage 1 (2nd stage)

Titan III

Solid Rocket Motors (SRMs) (1st stage)

Titan/Centaur Launcher

The 1975 launch window is much less favourable than either 1971 or 1973 would have been. It was reluctantly postponed following the 1972 budget cuts; so the plan to link the first soft-landing with the 200th anniversary of independence is particularly ambitious. The journey time of nearly a year, involves a flight of 440 million miles (700 million km), and arrival at Mars when the planet is about 206 million miles (330 million km) from earth on the other side of the sun. The 2 spacecraft, both including orbiter and lander sections, will be launched by Titan–Centaur rockets at least 10 days apart, and placed in a 115-mile (184-km) earth-parking orbit. Centaur re-ignites after 30 mins to propel the spacecraft on to a Mars trajectory. By the time Mars is reached, a one-way command from earth will take 20 mins; operations such as soft-landing must therefore be performed completely automatically by the onboard pre-programmed computer. As the Vikings approach Mars, their retro-rockets will be fired for nearly an hour to place them in highly elliptical orbits of 930 miles (1500 km) by 20,500 miles (33,000 km); they will then be tracked for at least 10 days, to check the pre-selected landing sites and ensure precise touch-downs. A recurrence of the gigantic dust-storm of 1972 is considered unlikely; but if there are any localized storms the spacecraft can wait in their parking orbits for up to 50 days. During this time the orbiter's solar panels provide electric power to both sections. After separation, and during the 6–13 mins between then and touchdown, the lander relays progress reports, and information on the Martian upper atmosphere and temperatures, via the orbiter to earth. At 20,000 ft (6000 m), a 50-ft parachute is deployed to slow the descent. Shortly afterwards, the aeroshell, a shield protecting the lander against the intense heat generated as it decelerates through the thin CO_2 atmosphere, is jettisoned, followed by the parachute about 1 mile (1·6 km) above the surface. For the remainder of the descent, at least 5 mins, the 3 terminal rocket engines gradually slow the lander for touchdown, cutting off as the 3 footpads touch the surface. Immediately on touchdown, the lander's computer determines its attitude on the surface to provide information for aligning the S-band transmitter/receiver, and the 2 35 W nuclear generators start recharging the lander's batteries for a mission aimed to last at least 90 days and possibly 1 year. The lander's life-detecting instruments include a soil sampler, equipped to detect and identify organic molecules, described as 'the building blocks of life'. 2 facsimile cameras, with a 360° panoramic scan, will take pictures in high-quality black-and-white and colour, and in the near infra-red. They will help in the selection of soil samples, and observe clouds and dust-

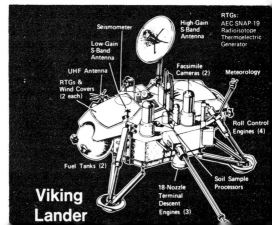

Viking Lander

Seismometer

Low-Gain S-Band Antenna

High-Gain S-Band Antenna

RTGs: AEC SNAP-19 Radioisotope Thermoelectric Generator

UHF Antenna

Facsimile Cameras (2)

Meteorology

RTGs & Wind Covers (2 each)

Roll Control Engines (4)

Fuel Tanks (2)

Soil Sample Processors

18-Nozzle Terminal Descent Engines (3)

storms as well as viewing the landscape. A seismometer will measure Marsquakes or meteoroid impacts. The lander can communicate with earth both directly and through the orbiter, but because of power limitations, its transmissions must be restricted to a few hours a day.

While the lander operates on the surface, its orbiter will circle overhead, observing the landing site with its 2 TV cameras, and making measurements to compare the landing-site conditions with the planet's overall characteristics. The 66 participating scientists from the US and other nations are hopeful that, even if Mars proves to be lifeless now, clues can be found indicating whether it was once the host for a rich variety of life that disappeared later in the planet's history. These hopes were revived by Mariner 9's discovery of volcanism and riverbeds. The cost of the Viking missions is estimated at between £312–£345 million ($750–$830 million), not including the £54 million ($132 million) for the 2 Titan–Centaurs.

Spacecraft Description Combined lift-off wt of orbiter and lander: 7500 lb (3400 kg). Ht 16 ft (4·9 m). Width 32 ft (9·8 m). Orbiter wt 5200 lb (2360 kg), including 3097 lb (1406 kg) propellant. Lander wt 2300 lb (1050 kg) including 307 lb (139 kg) propellant. Orbiter instruments weigh 144 lb (65 kg), while lander instruments weigh 133 lb (60 kg). Design of lifetime orbiters is 140 days, but hopefully they will operate as long as 2 years; orbital life, before they impact on Mars is over 50 years.

Future Missions None yet approved, but provisional plans are being made to land a Mars rover, somewhat similar to Russia's Lunokhod vehicle, during the 1977 launch window.

LAUNCHERS

Details of Soviet launch vehicles have never been fully published. They have been painstakingly assembled over a period of many years by Western observers. In the entries that follow I have used the identification system originated by Dr Charles Sheldon, of the US Library of Congress, combined with explanations given by Dr J. A. Pilkington, Director of Scarborough Planetarium.

By 1969 6 basic Soviet launch vehicles had been identified, classified as A, B, C, D, F and G; X was used for some launches which were difficult to classify. G was for 'Webb's Giant', named after the former NASA Administrator, who said in 1967 that Russia was developing a huge new launcher, bigger than Saturn 5. Reports that the development version blew up have never been confirmed, but it is now doubtful if it exists.

Added upper stages have been designated 1, 2, 3 etc; where there is doubt whether the stage used is 1 or 2, it is marked 1/2. The escape rocket, often a 4th stage, is labelled 'e'; manoeuvrable stages, 'm'; a re-entry rocket, 'r'.

Identification is further complicated because Soviet rockets are basically military, and known in the West by NATO code-names. These names are also included in the summary descriptions of the most frequently used combinations which follow:

A1 Vostok core + Luna stage. This early vehicle consists of the original Soviet ICBM, plus 4 tapered, strap-on boosters to make the first Vostok launcher, with a total lift-off thrust of 1,124,000 lb (509,830 kg). The 'cluster design'

('A') is regarded as 2 stages; the 3rd, Luna stage ('1') provides a single engine giving 199,000 lb (90,260 kg) thrust. It was so named because it first appeared with the Luna 1 launch, and subsequently in the Vostok, Electron and Meteor projects. Payload capacity of this combination is 10,400 lb (4720 kg) in low earth orbit, with launchpads at Tyuratum and Plesetsk.

A2 Vostok core+Venus stage. This standard vehicle consists of the Vostok core ('A') described above, plus a 2-engine 3rd stage, ('2') giving 309,000 lb (140,160 kg) thrust. It was so named because it first appeared in February 1961, to launch Venus 1, and was later used in the Voskhod and Soyuz projects. Payload capacity is 16,500 lb (7480 kg) in low earth orbit, with launchpads at Tyuratum and Plesetsk. For the Molniya, planetary, and later Luna programmes, an additional 'escape' stage was added (thus 'A2e').

B1 Sandal+Cosmos stage. This small, 2-stage launcher, first used for Cosmos 1 in March 1962, has as a 1st stage a modified Sandal IRBM (B), with thrust approx. 166,000 lb (74 tonnes). The single engine upper stage (1) has a thrust of approx. 24,000 lb (11 tonnes). The complete assembly is called the Cosmos launcher in Russia. It is used for small satellites akin to NASA's Explorers, weighing between 285–1000 lb (129–453 kg). Launchpads are at Kapustin Yar and Plesetsk.

C1 Skean+Restart stage. This intermediate launcher is believed to use as 1st stage a modified MRBM (C) of unknown thrust, with an unidentified upper stage (1), dubbed a restart stage because it is used to inject satellites into medium-height circular orbits. This requires 1st- and 2nd-

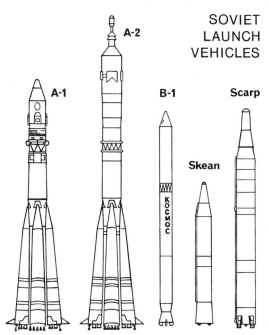

A-1

A-2

B-1

Scarp

Skean

КОСМОС

stage firing into elliptical orbit, coast to apogee, and then 2nd-stage re-ignition to circularize the orbit. C1 first appeared in August 1964 for Russia's first triple-launching (Cosmos 38–40); subsequent use has included Cosmos navigational satellites. Payload capacity up to 3200 lb (1451 kg) in low orbit, with launchpads at Tyuratum and Plesetsk.

D1 Proton core + Proton stage. This heavy launcher consists of a multi-engined 1st stage

(D) with 3,300,000 lb (1,496,880 kg) thrust, which can be used with or without an upper stage (1) and an escape (e) stage. The final stage was first used for the huge Proton satellite launchings beginning in July 1965, and later in the Zond programme. Payload capacity up to 50,000 lb (22,680 kg) for low earth orbit, 10,000 lb (4536 kg) to the moon, or 7000 lb (3175 kg) to the planets. Launchpads only at Tyuratum.

F1r   Scarp+FOBS stage. This military vehicle consists of either the 3-stage Scrag ICBM, or more probably the newer 2-stage Scarp ICBM (F1), with a re-entry FOBS (r). Both Scarp and Scrag are capable of orbital delivery of nuclear bombs to any point on earth. The final stage, dubbed FOBS stage (for 'Fractional Orbital Bombardment System'), because it has only been used in military FOBS tests, first appeared in September 1966 in the unannounced Cosmos U1 launch. Launchpads only at Tyuratum.

AUREOLE

Aureole 1   L. December 27, 1971, by C1 from Plesetsk. Wt ?660 lb (300 kg). Orbit 255 × 1553 miles (410 × 2500 km). Incl. 74°. The first Soviet–French satellite, launched under a new international project, 'Arcade'. It is studying the aurora borealis, or Polar lights, believed to be huge 'plasma explosions' far from earth, which change the ion composition of earth's upper atmosphere and affect radio conditions. Instruments were designed and supplied by Russia's Space Research Institute and France's Toulouse Centre for Study of Space Radiation. Orbital life of Aureole 1 is approx. 70 years.

COSMOS
Launch Rate

1962:	1–12	= 12		1968:	199–262	= 64
1963:	13–24	= 12		1969:	263–317	= 55
1964:	25–51	= 27		1970:	318–389	= 72
1965:	52–103	= 52		1971:	390–470	= 81
1966:	104–137	= 34		1972:	471–542	= 72
1967:	138–198	= 61		1973:	543	

History On July 10, 1972 the Soviet Union launched the 500th in the Cosmos series; since they began with Cosmos 1 on March 16, 1962 launches have averaged nearly 50 a year. At the time of writing this rate is being more than maintained. At least half of the series are military satellites, though this has never been admitted; in 1970, for instance, 57 of the 71 Cosmos launches were believed to be for military purposes. Their activities, ranging from the development of the ability to intercept and destroy other satellites, to the regular launching of pairs of overlapping 'spy' satellites, are described later. The variety of activity is reflected in the range of payloads; their weight varies from 285 lb (129 kg) to about 16,500 lb (7484 kg). The non-military Cosmos satellites continue to range over a wide area of research, and have achieved remarkable results; these results are usually announced months, or even years, after launch. Soviet scientists claim that these results are being widely used in the national economy, and nowadays the weekly Cosmos launch is accompanied by a brief announcement that it is 'carrying scientific instruments to continue space exploration under the programme announced earlier'. According to the *Soviet Encyclopaedia of Spaceflight* (1969) the research programme 'includes study of the concentration of charged particles, corpuscular fluxes, radio-wave propagation, distribution of the earth's radiation belt, cosmic rays, the earth's magnetic field, solar radiation, meteoric matter, cloud formations in the earth's atmosphere, solution of technological problems of spaceflight (including docking, atmospheric entry, effect of space factors, means of attitude control, life support, radiation protection etc), and flight testing of many structural elements and spaceborne systems'.

Cosmos numbers have always been used by Russia as a convenient way of concealing the inevitable failures that still accompany any space programme; interplanetary probes are only given designations in the Mars, Venus etc series when they are on course and working well. If they are unsuccessful they are merely given Cosmos numbers with the usual routine announcement about their launch. Cosmos 359 is an example. It should have been Venus 8, the second of a pair, but failed to achieve escape velocity after being launched from its earth-parking orbit.

The Cosmos series is launched from all 3 Soviet launch sites, at Tyuratum (Bykonur), Kapustin Yar, and Plesetsk. The last seems to be principally a military site; its existence was first made public in 1966, as a result of the tracking activities of Kettering Grammar School in England.

When Cosmos 500 was launched about 100 of the series was still in orbit, though not all were active. Some of the series at this stage were undoubtedly testing improved hatches for Soyuz spacecraft, following the catastrophic depressurization that caused the death of the Soyuz 11 crew.

Military Cosmos These are usually launched

at the rate of 2 a month, often in overlapping pairs. The launch rate rises during periods of tension, as was observed during clashes between Soviet and Chinese forces on 2 occasions in 1969. Reconnaissance satellites are mostly launched from Plesetsk and Tyuratum; the latter mostly have inclinations of 72°; others 52° and 65°. Until the summer of 1968 Russia's recoverable reconnaissance satellites remained in orbit for 8 days or less before being brought back for their film and electronic recordings to be processed. Starting with Cosmos 228, on June 21, 1968, they began to stay in orbit for 12 days or longer before ejecting their film packages; during 1971 the staytime of this type was frequently 14 days; presumably larger film packs made it possible to reduce the number of launches. Cosmos 251 in October 1968, and some later satellites such as 264 and 280, displayed limited manoeuvring capability to enable more precise coverage of the targets allotted to them.

Interceptor satellites, able to approach, inspect, and possibly destroy other satellites, began to appear in October 1968, with Cosmos 249 and 252. Western observers were able to study the new technique during October 20-30, 1970, by observing Cosmos 373, 374 and 375. As can be seen from the following list, 373 was probably sent up to play the part of an 'enemy satellite'; about 13 ft (3·9 m) long, with 7 ft (2·1 m) diameter, it was probably able to report back on 'miss' distances. Cosmos 374 went up 3 days later, and passed very close to 373 on its second orbit, and then exploded into over 16 pieces. Cosmos 375 was launched on October 30, and also passed very close to 373 on its second orbit, 230 mins after launch; it then exploded into 30 pieces. By the end of 1971 12 interceptor satellites had been iden-

Cosmos 144: early weather satellite

tified. By then Russia had succeeded in developing the ability to intercept and destroy satellites at relatively low level—much more difficult than at high level, because as the altitude decreases the target satellite moves faster in relation to a ground location. As can be seen on page 131, Cosmos 462 exploded on December 3, 1971, when approached by 459. During earlier interception tests target satellites had been destroyed at altitudes ranging between 360 and 550 miles (579 and 885 km). On this occasion the target was destroyed below 160 miles (257 km).

A possibly more aggressive use of space techniques began to be developed with 2 unnumbered launches (designation Cosmos U1 and U2) in September and November 1966. Later it was designated FOBS, as Western monitoring devices watched the tests with growing concern. The principle of this space bomb is that it can be fired into an orbit of 100 miles (160 km), but is slowed down by retro-rockets so that it re-enters and causes its nuclear warhead to fall on the target before completion of the first orbit. This provides a capability of attacking Western targets via the 'back door'—i.e. by travelling three-quarters of the way round the world via the South Pole, instead of by the shorter, more obvious North Pole route, which is monitored by BMEWS. In 1967 there were 9 FOBS tests, with Cosmos 139, 160, 169, 170, 171, 178, 179, 183 and 187. In the following 4 years, there were 9 more tests, at less frequent intervals, with none in the first half of 1972.

'Scarp' rockets, 113 ft 6 in. (34·5 m) long with 10 ft (3 m) dia., are used for FOB launches; the payload, which has been optically sighted from RAE, Farnborough, is estimated at 6 ft 6 in. (2 m) long and 4 ft (1·2 m) in dia. Undoubtedly some of the navigational satellites will be for use by nuclear submarines in targeting their ballistic missiles.

General Description Satellites placed in 49–56° orbits are usually cylindrical, about 6 ft (1·8 m) long by 3 ft 6 in. (1·0 m) dia., and 800 lb (360 kg) in weight. Military or reconnaissance satellites are also spheres, with an approx. wt of 7000 lb (3175 kg). The series as a whole, however, is so varied that they range from small, uninstrumented spheres to large vehicles like Cosmos 110, which carried 2 dogs with sufficient supplies to enable it to be recovered after remaining in orbit for 22 days. The orbits range from 90 miles (145 km) to 37,655 miles (60,000 km); the lifetime of the satellites vary from less than one orbit in the case of the military FOBS tests, to a possible 50,000 years. A complete, detailed list of Cosmos satellites would be too long for a book of this size; but typical examples, giving launch dates, orbits and inclinations, followed by a summary of the spacecraft's purpose and achievements, are listed below:

Cosmos 1 L. Mar 16, 1962 by B1 from Kapustin Yar. Wt ?440 lb (200 kg). Orbit 135 × 609 mi (217 × 980 km). Incl. 49°. At first classified as Sputnik 11; used radio methods to study structure of the ionosphere; decayed after 70 days. **Cosmos 2** L. Apr 6, 1962 by B1 from Kapustin Yar. Wt ?880 lb (400 kg). Orbit 132 × 969 mi (212 × 1560 km). Incl. 49°. At first classified as Sputnik 12. Returned data on radiation belts and cosmic rays and re-entered after 499 days. **Cosmos 3** L. Apr 24, 1962 by B1 from Kapustin Yar. Wt ?880 lb (400 kg). Orbit 142 × 447 mi (228 × 719 km). Incl. 49°. Returned radiation belt and cosmic ray data. Decayed after 176 days. **Cosmos 4** L. Apr 26, 1962 by A1 from Tyuratum. Wt ?8800 lb (4000 kg). Orbit 185 × 205 mi (298 × 330 km). Incl. 65°. The first spacecraft to be recovered; it was about 16·4 ft (5 m) long and 6·6 ft (2 m) in dia., and re-entered after 3 days. The first military satellite, since its task was to measure radiation before and after US nuclear tests. **Cosmos 97** L. Nov 26, 1965 by B1 from Kapustin Yar. Wt ?880 lb (400 kg). Orbit 137 × 1305 mi (220 × 2098 km). Incl. 49°. First experiment in measuring mazers; tested a molecular quantum generator, which makes it possible to communicate with, and control other spacecraft, and to send information great distances. Also checked aspects of the theory of relativity. Decayed after 492 days. **Cosmos 110** L. Feb 22, 1966 by A2 from Tyuratum. Wt ?8800 lb (4000 kg). Orbit 116 × 562 mi (186 × 904 km). Incl. 52°. Biological satellite, carrying dogs Veterok and Ugolyok, who were successfully recovered after 330 orbits in 22 days. **Cosmos 122** L. Jun 25, 1966 by A1/2 from Tyuratum. Orbit 388 × 388 mi (625 × 625 km). Incl. 65°. Meteorological satellite. Launch witnessed by General de Gaulle. Expected life 50 yrs. **Cosmos 144, 156, 184, 206** L. between Feb 1967 and Mar 1968 into circular 390-mi (628-km) orbits at 81° incl. with lifetime of 50–60 yrs. Part of Meteor system. **C. 144** and **156** carry equipment for TV and infra-red photography, providing pictures of cloud layers, and of snow and ice-fields for about 8% of earth's surface; also measure radiation streams reflected and emitted by earth and its atmosphere over about 20% of earth's surface on each orbit. The area scanned by one is reviewed by the second 6 hrs later. **C. 206** is about 20 mins behind **184**, so that forecasters can check weather received from the first. **Cosmos 166, 215** L. Jun 16, 1967 and Apr

19, 1968 by B1 from Kapustin Yar. Wt ?880 lb (400 kg) into orbits between 162 × 359 mi (260 × 577 km). Incl. 48°. Studied solar radiation. The second had 8 mirror telescopes, an X-ray telescope and 2 photometers to observe the radiation of hot stars in various wavebands—a first step, according to Pravda, towards placing a big telescope beyond the confines of the earth's atmosphere. Their orbits decayed after 130 and 72 days respectively. **Cosmos 186, 188** L. Oct 27 and 30, 1967 by A2 from Tyuratum. Wt ?13,200 lb (6000 kg). Orbits between 130 × 171 mi (209 × 276 km). Incl. 51°. Carried out world's first automatic docking, and Russia's first docking of any kind. **C. 186,** the 'active' craft, automatically manoeuvred to rendezvous and dock with **C. 188.** After 3½ hrs they were commanded to undock, and manoeuvred into different orbits. The mission, in Soyuz orbits, was clearly a rehearsal for manned flight. Both craft were recovered during the 4th day. **Cosmos 212 and 213** Achieved a second automatic docking on Apr 15, 1968, 5 months before the joint Soyuz 2 and 3 flights, but the latter did not succeed in docking. **Cosmos 187** L. Oct 28, 1967 by F1r from Tyuratum. Orbit 90 × 130 mi (145 × 209 km). Incl. 50°. Believed to be the first test of FOBS, which enables a nuclear warhead to be sent round the world, avoiding existing missile early warning systems. The warhead re-enters and descends on its target without completing 1 orbit. C. 187's weight was unknown; it was cylindrical, approx. 6·6 ft long × 3·3 ft dia. (2 m × 1 m). **Cosmos 243** L. Sep 23, 1968 by A1/2 from Tyuratum. Wt ?88800lb (4000 kg). Orbit 130 × 198 mi (209 × 319 km). Incl. 71°. Regarded by Soviet scientists as a landmark in the series; first satellite to study heat-ray

Cosmos 186/188: first Soviet docking

emissions from earth and its atmosphere. It enabled an Antarctic ice map showing temperature distributions around the world to be made. Registering heat radiation in this way enabled scientists to determine moisture content in the atmosphere, and to discover focal points of intensive precipitation concealed by thick clouds. Oceans were huge accumulators of solar energy, said Soviet scientists, which was emitted as evaporation heat; this heat 'fed' the cyclones which made the earth's weather. Cross-sections of water-surface temperatures in the Pacific from the Bering Sea to the Antarctic were 'mapped in tens of minutes'. C. 243 was probably recovered after 11 days. **Cosmos 248, 249, 252** L. Oct 19, Oct 20 and Nov 1, 1968 by F1m from Tyuratum into orbits ranging from 304 × 1353 mi (547 × 2175 km). Incl. 62°. Military reconnaissance satellites with ability to manoeuvre, with lives of 10, 100 and 200 yrs. **Cosmos 251, 264, 280** L. Oct 31, 1968, Jan 23 and Apr 23, 1969 from Tyuratum. Wt ?8800 lb (4000 kg). Orbits ranging from 123 to 205 mi (198 × 402 km). Incl. 65°, 70° and 51°. Military reconnaissance satellites, with manoeuvring capability for more precise target coverage. Cosmos 251 ejected a capsule about half its own weight after 12 days in orbit, and was recovered after 18 days. 264 and 280 ejected capsules after about 11 days and were themselves recovered on the 12th day. The capsules each remained in orbit for several days before decaying; their purpose is unclear though the recovered portions undoubtedly brought back film packages. There were 4 capsule ejections in 1969; 9 in 1970; and 24 in 1971. **Cosmos 261** L. Dec 20, 1968 by B1 from Plesetsk. Wt ?881 lb (400 kg). Orbit 135 × 416 mi (127 × 669 km). Incl. 71°. This paved the way for the Intercosmos programme. Bulgaria, Czechoslovakia, E. Germany, Hungary, Poland and Romania collaborated in experiments exploring air density in the upper atmosphere, and the nature of the Polar auroras. Decayed Feb 12, 1969. **Cosmos 262** L. Dec 26, 1968 by B1 from Kapustin Yar. Wt ?881 lb (400 kg). Orbit 163 × 508 mi (262 × 965 km). Incl. 48·5°. First satellite to study vacuum ultra-violet (VUV) and soft X-ray radiation (SX) from the stars, sun and earth's upper atmosphere. Carried 3 16-channel photometers. Results announced Oct 1969. Decayed after 4 months' operation, Jul 18, 1969. **Cosmos 336–343** L. Apr 25, 1970 by C1 from Plesetsk. Orbits from 816 to 966 mi (1313 × 1554 km). Incl. 74°. First octuple, or 8-satellite launch, each believed to be spheroid, weighing 88 lb (40 kg) and about 3 ft (1 m) long, 2·4 ft (0·8 m) dia. **Cosmos 359** L. Aug 22, 1970 by A2e from Tyuratum. Wt 2601 lb (1180 kg). Orbit 553 × 129 mi (889 × 208 km). Incl. 51°. Almost certainly intended to be Venus 8, but failed to achieve escape velocity. (Venus 7 was launched 5 days earlier.) **Cosmos 373** L. Oct 20, 1970 by F1m from Tyuratum. Orbit 338 × 293 mi (543 × 472 km). Incl. 62°. Target for satellite intercept system; probably about 13 ft (4 m) long, 7 ft (2 m) dia., containing devices to measure 'miss' distances. 10-yr life. (*See* below.) **Cosmos 374** L. Oct 23, 1970 by F1m from Tyuratum. Orbit 1331 × 324 mi (2142 × 521 km). Incl. 62°. Satellite intercept test; passed close to C. 373 on second orbit and exploded into 16 pieces. **Cosmos 375** L. Oct 30, 1970 by F1m from Tyuratum. Orbit 1305 × 326 mi (2100 × 524 km). Incl. 62°. Continued satellite intercept test; passed close to C. 373 on second orbit and exploded into 30 pieces. **Cosmos 394** L. Feb 9, 1971 by F1m from Plesetsk. Orbit 385 × 357 mi (619 × 574 km). Incl. 65·9°. Orbital intercept target for C.397; life 40 yrs. **Cosmos 397** L. Feb 25, 1971 by F1m from Tyuratum. Orbit 1440 × 369 mi (2317 × 593 km). Incl. 65·8°. Orbital intercept test; passed close to C.394. Life 150 yrs. **Cosmos 400** L. Mar 18, 1971 by F1m from Plesetsk. Orbit 625 × 612 mi (1006 × 903 km). Incl. 65·8°. Orbital intercept target for C.404. Life 1200 yrs. **Cosmos 404** L. Apr 3, 1971 by F1r from Tyuratum. Orbit 626 × 503 mi (1009 × 811 km). Incl. 65·1°. Orbital intercept attacker; passed close to C.400; did not explode and deorbited after about 5 orbits. **Cosmos 411–418** L. May 7, 1971 by C1 from Plesetsk. Second octuple launch; orbits 956 × 921 mi (1539 × 1482 km) approx. Wt 90 lb (41 kg). Incl. 74°. Navigational satellites; lifetimes 5000–10,000 yrs. **Cosmos 419** L. May 10, 1971 by Proton from Tyuratum. Orbit 90 × 99 mi (145 × 159 km). Incl. 51°. Attempted Mars probe; failed to leave earth orbit. Decayed after 2 days. **Cosmos 433** L. Aug 8, 1971 by F1r from Tyuratum. Orbit 161 × 99 mi (259 × 159 km). Incl. 49°. FOBS test; possibly recovered just short of 1 orbit. **Cosmos 444–451** L. Oct 13, 1971 by C1 from Plesetsk. Wt 90 lb (41 kg). Orbits 978 × 927 mi (1574 × 1492 km) approx. Incl. 74°. Third octuple launch; probably navigation satellites; said to be 'carrying radio systems for measuring elements of the orbit, and radio telemetric systems for relaying to earth data about the functioning of instruments and scientific equipment'. Life 6000–10,000 yrs. **Cosmos 459** L. Nov 29, 1971 by F1m from Plesetsk. Orbit 172 × 140 mi (277 × 266 km). Incl. 65°. Orbital intercept test; passive target for C.462; probably measured 'miss' distance. Life 28 days. **Cosmos 462** L. Dec 3, 1971 by F1m from Tyuratum. Orbit 1143 × 147 mi (1840 × 237 km). Incl. 65°. Caught up with C.459 on second orbit near Plesetsk and exploded into more than 16 pieces. **Cosmos 482** L. Mar 31, 1972 by Vostok from Tyuratum. Orbit 127 × 6090 mi (204 × 9800 km). Incl. 52°. Intended to be Venus 9, to accompany Venus 8, launched 4 days earlier; escape stage fired only partially; life 6 yrs. **Cosmos 496** L. Jun 26, 1972 by Proton from Tyuratum. Orbit 121 × 213 mi (195 × 343 km). Incl. 51°. Probably test of equipment for manned spaceflight; possibly redesigned Salyut/Soyuz hatch. Life 10 days. **Cosmos 504–511** L. Jul 20, 1972 by C1 from Plesetsk. Wt 88 lb (40 kg) approx. Orbits 898 × 931 mi (1446 × 1497 km) approx. Incl. 74°. Fourth octuple launch, 'carrying radio systems for precision measurement of orbital elements and radio-telemetric systems to transmit instrument readings back to earth'. Lifetimes 5000–10,000 years; rockets 20,000 years. **Cosmos 520** L. Sep 19, 1972 by F1r from Tyuratum. Orbit 141 × 416 mi (227 × 669 km). Incl. 62°. There was US speculation that this was intended as a high-altitude intercept, due to meet C. 516 head-on over Atlantic. If so, approach was not close enough to be successful.

ELEKTRON

Elektron 1 and 2 L. January 30, 1964 by A1 from Tyuratum. Wt 771 and 981 lb (350 and 445 kg). Orbits 244 × 4427 miles, and 274 × 42,246 miles (394 × 7126 km, and 441 × 67,988 km) Incl. 61°. Russia's first dual launch, the satellites being separated from the launch vehicle while the last stage was still firing. They studied the inner and outer zones of the Van Allen radiation belt and the earth's magnetic field, gathering information for radiation protection of manned spacecraft. Orbital life 200 and 30 years.

Elektron 3 and 4 L. July 11, 1964 by A1/2 from Tyuratum. Similar weights, orbits and missions. Orbital life 200 and 23 years.

INTERCOSMOS

History A series of international research satellites, aimed at expanding research opportunities in, and applying space techniques to the national economies of the 7 participating countries: Bulgaria, Hungary, German Democratic Republic, Poland, Romania, Czechoslovakia and USSR. First launch was Cosmos 261, December 20, 1968, from Plesetsk, which studied air density and Polar auroras. Subsequent launches have been from both Tyuratum and Kapustin Yar. Annual conferences are held, in which Cuba also participates, to review progress and plan future missions. Intercosmos launches apparently overlap the work of both Proton and Prognoz launches; defending this situation, when Intercosmos was launched, Academician Boris Petrov said that while the layman might think the flight programmes repeated themselves, the later flights were a continuation of earlier research with improved accuracy of measurements and greater capacity to handle the information.

Intercosmos launches average 2 per year. Usual weight is around 880 lb (400 kg); although No. 6, L. April 7, 1972, wt 9000 lb (4082 kg) into 126×159 miles (203×256 km) was a new, recoverable type. It was brought back after 4 days, with records of a 'space wanderer', with a pulse of 1000 million million electronvolts. The 9th launch, on April 19, 1973, was a Soviet–Polish occasion, and the satellite was named Intercosmos Copernicus-500, to celebrate the 500th anniversary of the birth of the Polish scientist. In an orbit of 125×964 miles (202×1551 km), with 48° incl., it spent 6 months studying the effects of solar radiation on the earth's atmosphere until re-entering.

LUNA

History Soviet scientists must have been planning lunar flights and exploration several years before their first satellite, Sputnik 1, was orbited in October 1957. Luna 1, 15 months later, was only Russia's fourth space launch. Details given below, and also under Lunokhod, show the Russian technique of tackling and solving each problem with a group of spacecraft, and then pausing for a year or two of research and development before going on to the next phase. Thus the 'fly-by' technique was mastered with 3 flights in 1959, culminating in the transmission by Luna 3 of the first historic pictures of the moon's far side. The next phase, preceded by a failure (Sputnik 25, on January 4, 1963) began the development of soft-landing techniques; though none appears to have been completely successful, much was learned from Lunas 4–8. Cosmos 60, on March 12, 1965, and Cosmos 111, March 1, 1966, were almost certainly 2 of only 5 complete failures in this long programme. The group of 5 Lunas launched in 1966 marked the biggest leap forward, achieving both soft-landing and orbiting techniques. The following entries reveal an interesting example of the Soviet policy of strict secrecy about long-term planning: it was only during Lunokhod 1's first activities on the lunar surface that we were told that Luna 12, 4 years earlier, had space-tested the electric motor for the robot's wheels. There were follow-up trials of the gears on Luna 14. There seemed to be a touch of desperation about Luna 15, which in July 1969 crashed on the lunar surface during an attempt to make an automatic recovery of lunar samples and get them back to earth a few hours before the

Luna 1: pre-launch assembly

Apollo 11 crew. This was the first time that Russia had tried a soft-landing from parking orbit; previous attempts had always been by direct flight—the system still employed in Martian and Venusian landings at the time of writing (mid-1973). The more sophisticated parking-orbit technique, which gives time for orbital corrections and makes pin-point touchdown possible, has always been preferred by the Americans, and was finally employed by Russia from Luna 16 onwards. (Luna 16 was apparently preceded by 2 launch failures, Cosmos 300 and 305, in September and October 1969.) When Lunokhod exploration began, Academician Boris Petrov claimed that robots could carry out such missions for less than one-tenth the cost of a similar manned flight. It seems certain that the Luna programme will include, in the very near future, Lunokhods exploring the moon's polar regions, and opera-

135

tions of the far side under the command of satellites, such as Luna 19 which was usable for more than a year.

Luna 1 L. Jan 2, 1959 by A1 from Tyuratum. Wt 797 lb (361 kg). Intended to impact on the moon, this was the world's first spacecraft to reach 'second cosmic velocity', or 25,000 mph (40,234 kph). Launch believed to be by 3-stage vehicle with total thrust of 580,000 lb (263,000 kg). The spherical craft, equipped with instruments for measuring radiation, etc, and its separated 3rd stage, both passed within 3700 mi (5955 km) of the moon, and went into solar orbit. Also named 'Mechta', it was only Russia's fourth space launch. **Luna 2** L. Sep 12, 1959, by A2 from Tyuratum. Wt 860 lb (390 kg). The first spacecraft to reach another celestial body; impacted E. of Sea of Serenity; carried pennants bearing USSR coat of arms, the hammer-and-sickle emblem. **Luna 3** L. Oct 4, 1959 by A1 from Tyuratum. Wt 613 lb (278·5 kg). First spacecraft to pass behind the moon and send back pictures of far side; placed on an elliptical earth orbit with apogee of 298,260 mi (480,000 km), so that without midcourse corrections, lunar gravity pulled it around the moon at a distance of about 3852 mi (6200 km). Equipped with a TV, processing and transmission system, it took an unannounced number of farside pictures which were sent back as Luna 3 moved back towards earth; 3 were published, including a composite full view of the far side, and 2 large mare were named Mare Moscovrae (Moscow Sea) and Mare Desiderii (Dream Sea). Decayed after 11 orbits totalling 177 days. **Luna 4** L. Apr 2, 1963 by A2e from Tyuratum. Wt 3135 lb (1422 kg). First of 5 spacecraft aimed at solving problems of soft-landing instrument containers. Contact lost as it missed moon by 5300 mi (8529 km), leaving it in 55,800 × 434,000-mi (89,782 × 692,300-km) earth orbit. **Luna 5** L. May 9, 1965 by A2e from Tyuratum. Wt 3254 lb (1476 kg). First soft-landing attempt. Flight lasted 82 hrs, and at 60 hrs Soviet scientists for first time made advance announcement of their plans. Retro-rocket, due to be fired about 40 mi (64 km) from surface, malfunctioned, and spacecraft impacted in Sea of Clouds 5 mins earlier than planned. **Luna 6** L. Jun 8, 1965 by A2e from Tyuratum. Wt 3179 lb (1442 kg). During midcourse correction, manoeuvre engine failed to switch off. Spacecraft missed the moon by nearly 100,000 mi (161,000 km), and passed into solar orbit. **Luna 7** L. Oct 4, 1965, by A2e from Tyuratum. Wt 3320 lb (1506 kg). After 86-hr flight, retro-rockets fired early; crashed in Ocean of Storms. **Luna 8**

Luna 2: first
spacecraft to
impact on moon

Luna 3 obtained
first photo of
Lunar far side

L. Dec 3, 1965 by A2e from Tyuratum. Wt 3422 lb (1552 kg). After 83-hr flight, retro-rockets fired late; crashed in Ocean of Storms. **Luna 9** L. Jan 31, 1966, by A2e from Tyuratum. Wt 3490 lb (1583 kg). First successful soft-landing, followed by first TV transmission from surface. After 79-hr flight the spacecraft ejected an egg-shaped 2-ft (0·6-m) dia. instrument capsule, weighing 220 lb (100 kg), as a probe touched the surface and switched off the retro-rocket. The capsule, weighted so that it rolled into an upright position, was stabilized on earth-command by 4 spring-ejected 'petals' to serve as legs. 3 panoramas of the lunar landscape, with different sun angles, on the eastern edge of the Ocean of Storms, were transmitted over a 3-day period. **Luna 10** L. Mar 31, 1966 by A2e from Tyuratum. Wt 3527 lb (1600 kg). First lunar satellite; wt was 540 lb (245 kg) when it was fired into 217 × 632-mi (349 × 1017-km) lunar orbit, with 71° incl. In addition to broadcasting the 'Internationale' several times, it studied lunar surface radiation, magnetic field intensity etc. Communications were maintained for 2 months and 460 orbits, providing opportunities for tracking the strength and variation of lunar gravitation. **Luna 11** L. Aug 24, 1966 by A2e from Tyuratum. Wt 3616 lb (1640 kg). Second, heavier lunar satellite. Placed in 99 × 746-mi (159 × 1200-km) lunar orbit with 27° incl. Possibly carried TV system which failed to operate. Data was received during 277 orbits until Oct 1, 1966. **Luna 12** L. Oct 22, 1966 by A2e from Tyuratum. Wt 3616 lb (1640 kg). Third lunar satellite; placed in 62 × 1081-mi (100 × 1739-km) orbit. A TV system transmitted large-scale pictures of the Sea of Rains surface and Crater Aristarchus areas, showing craters as small as 49 ft (15 m) across. Tested electric motor for Lunokhod's wheels. Communications continued for 602 orbits until Jan 19, 1967. **Luna 13** L. Dec 21, 1966 by A2e from Tyuratum. Wt ?3490 lb (1583 kg). Second successful soft-landing. Capsule was again bounced on to the Ocean of Storms. In addition to sending back panoramic views, 2 Meccano-like arms 5 ft (1·5 m) long, were extended, which measured soil density and surface radioactivity; communications lasted for 6 days from landing. **Luna 14** L. Apr 7, 1968 by A2e from Tyuratum. Fourth Soviet lunar satellite, placed in 99 × 540-mi (160 × 870-km) orbit, with 42° incl. Studied moon's gravitational field, and 'stability of radio signals sent to spacecraft at different locations in respect to the moon'; made further tests of geared electric motor for Lunokhod's wheels. **Luna 15** L. Jul 13, 1969 by D1e from Tyuratum. Wt ?4000 lb (1814 kg). A bold but unsuccessful attempt to obtain lunar samples and return them to earth a few hours before America's first men on

138

Luna 9, first soft-lander: model in Moscow Exhibition

the moon (Apollo 11) could do so. Launched $3\frac{1}{2}$ days before Apollo 11, Luna 15 was placed in lunar orbit, and remained there during the Apollo flight and first manned landing. 2 orbital changes were made, however, and American scientists, concerned about possible conflict of radio frequencies, asked Frank Borman, who had recently visited Moscow, to seek assurances from Russia that there would be no radio conflict. Such assurances were quickly given. 2 hrs before Apollo 11 was due to lift-off from the Sea of Tranquillity, Luna 15, on its 52nd revolution, began descent manoeuvres, apparently aimed at the Sea of Crises. But Russia's first attempt at an automatic landing from lunar orbit went wrong, and crashed at the end of a 4-min descent. Few details were officially announced, and it was more than a year later before the Luna 16 flight revealed what had been intended. **Luna 16** L. Sep 12, 1970 by D1e from

Tyuratum. Wt ?4000 lb (1814 kg). First recovery of lunar soil by automatic spacecraft. Luna 16 was placed in a 68-mi (110-km) orbit with 70° incl., corrected prior to landing to 68 × 9 mi (110 × 15 km) with 71° incl. On Sep 20 Russia's Long Range Space Communications Centre gave the command to fire the descent engine. At 66 ft (20 m) above the surface this was switched off, and final touchdown on the Sea of Fertility, 0·9 mi (1·5 km) from target point, was controlled by 2 vernier engines. Luna 16 consisted of landing and ascent stages; as in the case of America's Lunar Module, landing or descent stage served as a launch platform for the ascent stage. On earth command, an automatic drilling rig was deployed; with a 3-ft (0·9-m) reach, it was capable of penetrating the surface just over 1 ft (30 cm). About 100 gr (less than 4 oz), were lifted by the drill into the loading hatch of a spherical capsule at the top of the ascent stage, which was then hermetically sealed. After 26½ hrs on the surface, the ascent stage was launched on a ballistic trajectory back to earth; it consisted of the lunar soil container, an instrument compartment, and a parachute compartment, containing braking and main parachutes and 2 gas-filled balloons, presumably in case of descent on return flight; 3 hrs before re-entry, the instrument compartment was jettisoned. The sample capsule's transmitters enabled it to be located and recovered in Kazakhstan on Sep 24. **Luna 17** L. Nov 10, 1970 by D1e from Tyuratum. Wt ?4000 lb (1814 kg). Carrying the first moon robot, Lunokhod 1 was placed in an initial circular 52-mi (84-km) orbit with 141° incl.; the following day, the perilune was lowered to 11·8 mi (19 km). On Nov 17 a successful soft-landing was made in a shallow crater in the north-western Sea of Rains (Mare Imbrium). After checks by TV cameras that there were no boulder obstructions, one of 2 alternative ramps was lowered by commands from Russia's Deep Space Communications Centre, and Lunokhod 1 rolled down on to the surface. It proved to be the Soviet Union's greatest technical space success. Full details are given under Lunokhod. **Luna 18** L. Sep 2, 1971 by D1e from Tyuratum. Wt ?4000 lb (1814 kg). Intended as a soft-lander, most probably with an ascent stage for a second soil recovery, this was placed in a 62-mi (100-km) circular lunar orbit with 35° incl. After 54 orbits taking 4½ days (twice as long as Luna 16 and 17), an attempt was made to land in highland terrain in the Sea of Fertility. Communications ceased shortly after earth command had started the descent engine, probably as a result of impact. Novosti reported that the landing 'in difficult topographic conditions had been unlucky', the first time, within the author's knowledge, that Russia had publicly admitted a space failure. **Luna 19** L. Sep 28, 1971 by D1e

Luna 12 orbiter automatic station: 1 gas-containers; 2 TV; 3 thermal radiator; 4 radiometer; 5 instrument compartment; 6 chemical battery; 7 orientation system; 8 antenna; 9 electronic unit; 10 control jets; 11 retro-engine

Luna 13, showing petal-type antennae

from Tyuratum. Wt ?4000 lb (1814 kg). 5th Soviet lunar satellite, intially placed in circular 87-mi (140-km) orbit with 40° incl. Communications continued for over a year; on Oct 3, 1972 it was stated the experiment was nearing completion. By then Luna 19 had made more than 4000 lunar orbits, and over 1000 communication sessions had been held. A systematic study was made of the moon's gravitational field and the effect of its mascons, coupled with TV pictures of the surface. 10 powerful solar flares were observed. Lunar radiation was compared with similar measurements made by Mars 2 and 3 in Martian orbit and by Venus 7 and 8, and Prognoz 1 and 2. **Luna 20** L. Feb 14, 1972 by D1e from Tyuratum. Wt ?4000 lb (1814 kg). Second successful soil recovery. Placed in initial circular lunar orbit of 62 mi (100 km) at 65° incl. This was lowered 1 day later to 62 × 13 mi (100 × 21 km). After site selection photography, the descent engine was fired for 4 mins 27 secs, and $7\frac{1}{2}$ days after launch a safe landing was made on a mountainous isthmus pock-marked with large craters, south of the Sea of Crises, and on the Sea of Fertility's extreme NE; it was 74 mi (120 km) N of Luna 16's sampling point. A 'photo-telemetric device' relayed to earth pictures of the surface, and from these a site with 'a grey cloudy structure' was chosen from which to take samples. A rotary-percussion drill, of improved design as a result of Luna 16, drilled into the rock; it sank quickly to a depth of 4–6 in. (100–150 mm); then because of the rock's hardness, drilling had to be done in stages, with intervals so that the drill did not overheat. The samples were then lifted into the return capsule on Luna 20's ascent stage, which was fired back to earth on a ballistic trajectory 1 day after the touchdown. It was recovered with some difficulty on an island in the River Karakingir, 25 mi (40 km) NW of Dzezkazgan, in Kazakhstan, after landing in a blizzard on February 25. Newsmen were present 2 days later, when the contents of the hollow drill brought back from the moon were poured out. The ash was light-grey compared with the dark, 'black slate colour with a metallic glitter', of the Luna 16 samples. Scientists expected these samples to be about 1000 million years older than the Luna 16 samples, which were estimated at 3000–5000 million years old. **Luna 21** L. Jan 8, 1973, by D1e from Tyuratum. Wt ? over 4000 lb (1814 kg). Carried Lunokhod 2 to the moon; during translunar flight when false telemetry signals almost aborted the mission, there was also, apparently, a problem with the Lunokhod's power supply, and the rover's solar batteries were exposed to sunlight during most of the translunar coast. Descent was made on the 41st lunar orbit, from a height of 10 mi (16 km), and Lunokhod 2, wt 1850 lb

Luna 16 returning with lunar soil

(850 kg)—183 lb (84 kg) heavier than Lunokhod 1—was placed on the eastern edge of the Sea of Serenity. Touchdown was only 110 mi (180 km) from the Apollo 17 landing point. (*See* Lunokhod 2.)

LUNOKHOD, SOVIET 'MOONWALKER'

Wt: 1667 lb (756 kg)
Wheelbase: 64 in. (160 cm)
Dia. of wheel: 20 in. (51 cm)
Lid: 84 in. (215 cm)
Overall length of wheels: 87 in. (222 cm)

General Description Lunokhod consists of a circular instrument compartment, 7 ft (2·1 m) dia., mounted on an 8-wheel chassis. The 8 spoked, wire-mesh wheels are in 4 pairs; each has an independent electric motor; if any wheel seizes up, or gets stuck, a powder charge can be fired to break the drive shaft, enabling the wheel to become a passive roller; movement should still be possible with only 2 wheels on each side operational. Sensors provide automatic braking,

143

overriding earth commands, if slope angles become so steep that there is a danger of overturning. Disc brakes hold the craft at rest on slopes and during lunar nights. There are 2 forward speeds, and possibly 2 in reverse.

The instrument compartment, made of magnesium alloy, is designed so that it warms up easily and releases heat slowly, thus enabling it to survive lunar nights which are equivalent to 14 earth-nights. When not in use, the lid is kept tightly shut. During the lunar nights the instruments, cameras, etc are kept warm and operational despite outside temperatures of $-150°C$ by the circulation of heated gas. The instrument compartment contains radio transmitters and receivers, remote control, electric power and heat control systems, electronic-transformer units, and 2 TV systems. One of these, transmitting a frame every 3 to 20 secs, enables the operators to monitor its progress. The second TV system, to obtain panoramic pictures of the locality, the horizon, sun and earth, consists of 4 identical telephoto cameras; 3 look to the sides and rear; the fourth, mounted alongside the low-rate camera, looks through forward portholes. The fore and aft cameras have a combined range of 30° horizontally and 360° vertically, the side cameras 180° horizontally and 30° vertically.

Lunokhod 1 The first 'moonwalker' was landed on the Sea of Rains (Mare Imbrium) by Luna 17 on November 17, 1970. Its expected life was 90 days; but it continued to operate for 11 months. During that time, it travelled 34,500 ft (10,540 m) photographing an area of more than 861,000 sq ft (80,000 sq m). Over 200 panoramic pictures and 20,000 separate photographs were returned; physical and mechanical soil analyses were carried out at 500 locations and chemical analyses at 25. When at last Lunokhod's equipment froze during the 11th lunar night following exhaustion of its isotopic fuel, the vehicle had been placed so that its French laser-reflector could continue to be used indefinitely for earth–moon measurements. Academician Blagonravov said its performance had surpassed all expectations, and that Lunokhod could, in principle, make a 'round-the-moon' journey—though this would presumably involve the use of a lunar satellite for communications and control on the far side. Soviet scientists pointed out that Lunokhod was specifically designed for the moon; planetary robots would necessitate different designs, involving walking, crawling or hopping, according to conditions. In the case of Mars, with its atmosphere saturated with carbon dioxide, frictional surfaces would probably be unsuitable, because they would wear off much too fast; and the great distances from earth would mean revision of the remote-control methods employed for Lunokhod, which provided great difficulties themselves. Radio signals from earth to moon take $1 \cdot 3$ secs; from earth to Mars they take 14 secs. A 'Marsokhod', therefore, would have to be self-controlled by means of an onboard computer.

Operating Technique Lunokhod 1 was driven by a 5-man team—commander, driver, navigator, systems engineer and radio operator—working in the Deep Space Communications Centre, believed to be near Moscow. The need for co-ordinating their efforts in the early days was said to put great psychological and physical strain upon them; the technique was reported to be so difficult that it beat many highly experienced drivers and pilots. This was mainly because of the time-delays resulting from sending commands, and waiting to observe the response, over a distance of 240,000 miles (386,240 km). They had to remember that the robot was already several yards ahead of the slow-scan TV picture they were watching; and that if there were rocks or boulders to be avoided, it would have moved on still further before the signals to take avoiding action would reach it. However, the robot's automatic 'stop' system worked so well that throughout its 11-month life it never did overturn. The first task at the beginning of each lunar day was to open the lid to enable the solar cells on the lid's inside to start recharging the depleted batteries. When being parked at the end of the previous lunar day, it had been carefully turned towards the east, to obtain the maximum intensity from the sun's rays as soon as the lid was opened. Soon afterwards it became necessary to keep the upper part of the hull cool; a mirror-like surface reflected into space as much heat in an hour as a man engaged in heavy manual labour

Lunokhod: small wheel measures distance travelled

generated. As soon as movement began, telemetry poured in separately from each of the 8 wheels, so that action could be taken immediately if any one of them showed signs of over-heating or freezing. The revolutions of each wheel were also counted separately to check whether it was skidding or not. Initially they were the main source of anxiety; it was thought they would build up charges of static electricity because of the vacuum conditions on the lunar surface; without air it was feared that it would be impossible to run off the electricity, with the result that particles of soil would cling to the wheels, and ultimately clog them and make the robot inoperable. The fear was unjustified; in practice the soil fell off like volcanic sand—a factor which should make it possible to simplify future Lunokhods. Orientation proved another problem at first; on the third lunar day, scientists became excited by a big rock, estimated to be 492 ft (150 m) away. They asked the crew to move up close to the boulder; but when they did so it was nowhere to be seen. It turned out to be only a small stone, which on the screen appeared to be a huge rock. Attempts to steer Lunokhod closer to it, resulted in its being passed and left further behind. It was soon found that TV pictures of the wheel ruts provided a useful distance scale for navigators and drivers, as well as information for scientists about the physical and mechanical properties of the lunar soil. By the end of the second lunar day, the robot was climbing slopes with gradients of 23°, weaving, turning and zig-zagging to avoid steep craters and large boulders, and being sent with confidence through shallow craters 492 ft (150 m) across. Probably its most successful instrument was RIFMA (Roentgen Isotopic Fluorescent Method of Analysis), able to analyse the chemical composition of lunar rocks by sending out a stream of electrons, or X-rays. Under the action of the X-rays, the atoms of the rocks rearrange their electronic shells, and send back an X-radiation of their own. Measurements of this response show how much aluminium, silicon, magnesium, potassium, calcium, iron and titanium is contained in the rocks. On December 12, 1970 RIFMA reported a solar flare, which, according to Pravda, would have created hazardous radioactivity for men on the moon.

Lunokhod 2   Landed on the eastern edge of the Sea of Serenity in the Le Monnier crater on January 16, 1973, Lunokhod 2 appears to have ceased operating abruptly early in the 5th lunar day, which began on May 8. But though it operated for only half as long as its predecessor, it was much more active; its 9th, pedometer, wheel recorded that it had covered 23 miles (37 km), $3\frac{1}{2}$ times the distance; 86 panoramic pictures and 80,000 TV pictures of the lunar surface were transmitted. The Le Monnier landing site was chosen to provide information about a transitional 'sea/continental' area, and because there is a 10-mile (16-km) tectonic fault nearby. Improved equipment and additional instruments added nearly 220 lb (100 kg) to Lunokhod 2's weight. Control and communication systems had been improved, and one of the panoramic TV cameras, sited in the vehicle's centre, had been raised to improve picture quality and reduce time taken to relay signals back to earth, thus in turn giving the crew better control.

Equipment carried included a 'corner reflector' provided by French engineers, to continue the Franco–Soviet collaboration programme with laser-reflectors begun with Luna 17. Measurements made from earth and circumlunar orbit had already established that the visible side of the moon averages 1·2 miles (2 km) lower than the middle-radius, and that the far side is 6 miles (9½ km) higher. Additional measurements by means of the French corner reflector were aimed at filling in lunar contours with more precision, checking the theory that its continents, like those on earth, are subject to 'drift' etc. French laser-ranging staff were present at Mt Lock on January 26 . when the University of Texas McDonald Observatory succeeded in locating the reflector with a powerful laser attached to a 107-in (2·7-m) telescope. The reflected signal strength was comparable with that from the Apollo 15 array, the largest of the NASA arrays deposited by Apollos 11, 14 and 15.

First TV pictures showed that Lunokhod 2 was standing on an even plain between 2 small craters, with the low peaks of the Taurus mountains about 3·7 miles (6 km) away. Owing to the high position of the sun, the crew at the Space Communications Centre at first had difficulty in maintaining control; during the third session there was a near collision with the Luna 21 landing stage; the 2 vehicles were less than 13 ft (4 m) apart when Lunokhod 2 was stopped. Its speed was more than double that of Lunokhod 1, and operating it over the rising, terraced ground towards the Taurus mountains was described as 'very taxing' for the crew. They had to undergo medical checks, and eyesight and hearing reaction tests before each session, and wore sensors just as if they were themselves operating in space. First chemical tests of the regolith made by the Rifma instrument suggested that its composition was much the same as had been found on the Sea of Rains; the landing area generally, however, was more rugged, with more craters and boulders. During the first lunar night, the temperature at the end of the probe fell to a record −183°C, though the wheel temperatures never fell below 128°C.

During its second lunar day, occasionally sinking to the hubs of its wheels in loose rock, Lunokhod 2 moved steadily towards the Taurus mountains, taking magnetic measurements on the way. (Unlike the earth, the moon has no general magnetic field, probably because it has no liquid core; but the lunar seas and mainlands have local magnetic fields.) The robot zig-zagged up the mountain slopes, wheel-slip sometimes registering 80%. From a 1300-ft (400-m) peak, views were obtained of the opposite shore of the Le Monnier bay, and of the main mountain peaks 34–37 miles (55–60 km) away. It also sent back an unexpected view of earth, appearing as a 'thin sickle' in the local sky.

During the third day, and throughout the fourth, Lunokhod 2 was worked along the narrow precipice, or tectonal abyss, 100–160 ft (30–50 m) deep, and 980–1300 ft (300–400 m) wide. Moving cautiously along the westward edge, and then in the reverse direction along the eastern slope, it established that it was solid basalt rock which had been split by tectonic forces. On May 9 it was reported that movement had begun, at the start of the fifth day, away from the abyss in a north-easterly direction. There were no further reports until, on June 3, it was announced that the robot's research programme was completed. No indication was given as to whether there had been a mechanical failure, or whether it had been lost or damaged by falling into a crater or fissure.

Perhaps Lunokhod 2's most interesting discovery was that while the moon almost certainly has admirable conditions for astronomic observations during the lunar night, conditions are not nearly so good as had always been imagined during the day. An astrophotometer (an electron telescope without lens), registered glow in the lunar sky, both in the visible and ultra-violet bands of the spectrum, to establish whether the lunar 'atmosphere' contained cosmic dust which could affect observations from the surface. The after-sunset glow proved much higher than expected—10–15 times brighter than on earth, suggesting that the lunar sky is surrounded with a swarm of dust particles, with the effect of scattering solar light. This high luminosity would make daytime astronomy 'ineffective'.

MARS

History Details of Soviet probes and spacecraft to Mars, given below, clearly show that Russian scientists were well advanced with plans for interplanetary exploration long before the launch of Sputnik 1. First official Martian launches were Sputniks 22 and 24 in late 1962, preceded by Sputniks 19–21 aimed at Venus, and followed by Sputnik 25 aimed at the moon. As the launch table shows, it is Soviet practice to make 3 launches on each occasion; in 1962 none was successful, although much was learned from the only one successfully placed on course towards the Red Planet. In mid-1971 2 out of 3 reached Mars. Experience with Martian spacecraft led Soviet scientists to conclude that radio communications were possible over distances of 620 million miles (1000 million km); eventually they would be possible throughout the entire solar system, even though the signals reaching earth were so weak as to be measured in thousandths of a microvolt. Experience with Mars 2 and 3, and also Lunokhods 1 and 2 on the moon was expected to be used to develop 'Planetokhods' for robot exploration of the Martian surface in the 1973 or 1975 launch 'windows'; these Planetokhods would have to be largely operated, not from earth, but by onboard sensors, because of the time required for signals to be transmitted and obeyed.

Flight Techniques to Mars Mars is much further from earth than Venus, and is also further from the sun. Mean distance from the sun is 141 million miles (228 million km); in its larger orbit Mars moves more slowly than earth, so that its 'year' equals 687 of our days. With a diameter of 4217 miles (6786 km), Mars is about twice the size of the moon, and half the size of earth. Every second year, the 2 planets come within about 35 million miles (56 million km) of each other, providing launch 'windows' approximately every 25 months. According to the initial velocity given to the spacecraft, and the velocity chosen, Mars can be reached from earth in periods varying from 259 to 105 days. Faster journeys, however, mean higher fuel consumption by the launch rocket in order to achieve the great velocity; and this has the result of reducing the payload, or weight, of the spacecraft. In May 1971 the 2 planets were so favourably placed—factors such as the relative inclinations of the orbital planes of the 2 bodies providing good radio communications are part of it—that it was possible to put spacecraft on suitable trajectories with a velocity only a few hundred metres above the minimum escape speed. The flights of Mars 2 and 3 occupied respectively 192 and 188 days. Such favourable conditions will not recur until 1986. This explains why Russia prepares triple launches, and the Americans usually double launches. It was no coincidence that Mars 2 and 3 were in Martian orbit at the same time as America's Mariner 9. More details about the Red Planet itself may be found under the Mariner entry (pages 60–69).

It should be noted that the Soviet technique, at least in these early landing attempts, is to separate the landing craft for a direct approach before the parent craft is placed in Martian orbit. The US preference is first to place the whole vehicle in parking orbit, thus providing time for adjustments before the descent attempt is made. In the case of Mars 2 and 3 it meant that the landings had to go ahead despite the dust-storm raging on the surface; there was no possibility of holding the mission in Martian orbit, though presumably it was possible to adjust the targeted landing site during the final, pre-separation manoeuvre. It should also be noted that a Martian landing presents technical and design problems which are the exact opposite of those encountered on Venusian landings. Atmospheric pressure on the Venusian surface is about 100 times greater than on earth; on the Martian surface pressure is about 100 times less. Thus, while the dense Venusian atmosphere will quickly decelerate an approaching spacecraft, on Mars, if the approach angle is not exactly right, it will either pass through the thin atmosphere into space without slowing down, or enter too fast to enable the parachute to operate.

Spacecraft Description Photographs released shortly after Mars 2 and 3 had been placed in Martian orbit showed that they were about 12 ft (3·6 m) in height, and that more than half the launch weight consisted of fuel. The main hypergolic fuel tanks provided the central cylinder,

Mars 3: landing capsule at top

around which the spacecraft was built. The S-band, directional antenna dish, and 2 deployable solar panels were attached. A doughnut-shaped compartment, or toroid, at the base of the structure, containing electronics and instruments, completely enclosed the engine. The landing craft, protected by a heatshield shaped like an inverted saucer, consisted of a sphere, resting in a toroid, with its engine nozzle projecting through it. The latter contained the parachute system; the sphere was described as the 'automatic Mars station'. The heatshield provided maximum aerodynamic braking as the spacecraft entered the thin Martian atmosphere; it was discarded shortly before the drogue and main parachutes were explosively deployed; the parachutes were jettisoned at about 100 ft (30 m) to reduce final landing weight just before a solid-fuel braking engine was fired by a radar altimeter. Separation of the lander, followed by the placing of the orbiter into an elliptical path around Mars, was carried out by an onboard navigation/computer complex, updated by earth commands. The Mars 3 lander carried mass spectrometers for chemical analysis of the atmosphere, and equipment to analyse the chemical and physical content of the surface soil, wind anemometer, thermometer and pressure gauge and at least one TV camera. Data was to have been relayed via the orbiters, but only 20 secs of signals were obtained. Each orbiter carried 2 cameras, 1 wide-angled, 1 with a telescopic 4° narrow-angle lens. The photographs were automatically developed and the image transmitted directly to earth by the 80 in. (200 cm) S-band aerial, but the system was apparently less successful than the US Mariner system, which breaks down the pictures into coded dots which are then reconstructed on earth by computer. Other orbiter instruments provided ultra-violet, infra-red and optical examination of Mars and its atmosphere; by using infra-red sensors to measure the thickness of the carbon-dioxide atmosphere, variations in surface height could be found, and a relief map built up. Surface colouring could also be defined. Mars 3 also carried a French experiment, Stereo 1, to collect solar emission data for comparison with similar measurements on earth. With both US and Soviet spacecraft in Martian orbit at the same time, a teleprinter link was established between NASA's Jet Propulsion Laboratory at Pasadena, California, and the Soviet Co-ordinating and Computing Centre, for the exchange of information; Soviet information supplied over this link was not allowed to be published, but, apparently, merely duplicated Russian official statements.

Sputnik 22 L. October 24, 1962 by A2e from Tyuratam. Wt 1970 lb (893·5 kg). From an initial earth orbit of 112 × 252 miles (180 × 405 km), with 64° incl., the spacecraft and final rocket stage blew up when being accelerated to escape velocity. The debris passed within range of the West's BMEWS radars, which indicated a possible ICBM attack—especially alarming at that moment, occurring as it did during the Cuban missile crisis amid maximum East–West tension. But BMEWS computers, which assess trajectory and impact points, proved the alarm to be false within a few seconds.

Mars 1 L. November 1, 1962 by A2e from Tyuratam. Wt 1970 lb (893·5 kg). World's first Mars probe, successfully launched from earth orbit similar to Sputnik 22. Mars 1 was 10 ft 10 in. (3·3 m) long; with solar panels and radiators deployed following trans-Martian insertion, width was 13 ft 2 in. (4 m). It consisted of 2 hermetically sealed compartments, 'orbital and planetary'. The orbital section contained midcourse rocket engine, solar panels, high- and low-gain antennae, etc; the planetary section contained a photo-TV system and other instruments for studying the Martian surface during the period of closest approach. The Soviet Deep Space Tracking Station at Yevpatoriya in the Crimea held 61 communication sessions, first at 2-day, then at 5-day intervals until March 21, 1963; contact was then lost, because the craft's antenna could no longer be pointed towards earth, following a fault in the orientation system. At that point the craft had travelled 65·8 million miles (106 million km) from earth. It should have passed Mars 3 months later at a distance of between 620 and 6700 miles (998 and 10,783 km).

Sputnik 24 L. November 4, 1962 by A2e from Tyuratum. Orbit 122 × 367 miles (196 × 590 km). Incl. 64°. This too disintegrated during an attempt to place it on course for Mars from earth parking orbit; 5 major pieces of debris were tracked by BMEWS.

Cosmos 419 L. May 10, 1971 by D1e from Tyuratum. Wt ?22,046 lb (10,000 kg). First use of Proton launcher for a planetary mission. First of 3 craft intended for Mars; placed in 90 × 99-mile (145 × 159-km) earth orbit, but failed to separate from 4th stage and decayed in 2 days.

Mars 2 L. May 19, 1971 by D1e from Tyuratum Wt 10,251 lb (4650 kg). Reached Mars on November 27, after 192-day flight, during which 3 mid-

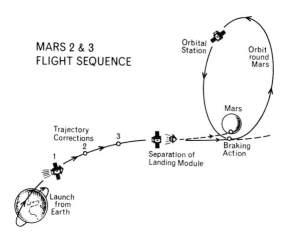

MARS 2 & 3
FLIGHT SEQUENCE

Orbital Station

Orbit round Mars

Mars

Trajectory Corrections

Separation of Landing Module

Braking Action

Launch from Earth

course corrections were made. Following the final manoeuvre, a landing capsule was separated from the orbiter, and made a first, unsuccessful attempt to soft-land. Few details were given, except that it had delivered a pennant with the USSR hammer-and-sickle emblem to the Martian surface. It was presumably destroyed on impact. After the lander was released, the orbiter's retro-rockets were fired; it became the second artificial Martian satellite, with an 18-hr, 860 × 15,500-mile (1380×25,000-km) orbit, at 48° incl. Within a few days Mars 2 established that at least nine-tenths of the Martian atmosphere consisted of carbon-dioxide; there was almost no nitrogen, and water vapour was very scarce.

Mars 3 L. May 28, 1971 by D1e from Tyuratum. Wt 10,251 lb (4650 kg). Reached Mars 5 days after its predecessor, on December 2. About $4\frac{1}{2}$ hrs before atmospheric entry the Martian lander was automatically separated from the parent craft, and placed on a shallow approach angle for a 3-min. descent. Aerodynamic braking on the saucer-shaped heatshield began at a speed of 13,400 mph (21,600 kph); the speed was still supersonic when the drogue parachute was explosively deployed. After main parachute deployment, the heatshield was jettisoned and radio antennae deployed. The soft-landing engine was switched on at about 98 ft (30 m) and the parachute fired to one side by another jet engine to prevent its canopy covering the landing capsule. 90 secs after touchdown, a timing device switched on the video-transmitter. For 20 secs 'a small part of a panoramic view' was transmitted by its TV camera, but in any case surface details were probably obscured by the dust storm then in progress. It was probably a combination of a heavy-landing and the dust-storm that overwhelmed the capsule. It had touched down in a pale-coloured region in the southern hemisphere, between Electris and Phaetontis (45°S, 158°W), in a rounded hollow about 930 miles (1500 km) across. A few days after the landing Soviet scientists said they assumed the surface was covered by 'something resembling the sands of the earth'; the reason for its changing colour during the Martian seasons was still unclear. After releasing the lander, the orbiter stage of Mars 3 fired its retro-rockets and went into an 11-day, 930 × 124,270-mile (1500 × 20,000-km) orbit, from which it sent back data for about 3 months.

Mars 4 and 5 L. July 21 and 25, 1973 from Tyuratum. The start of Russia's most ambitious onslaught on Mars, with 2 pairs of spacecraft (*see* below) being launched within 3 weeks—the first time a complete series had been successfully launched. Due to arrive in late January 1974, to 'explore Mars and its surrounding space'.

Mars 6 and 7 L. August 5 and 9, 1973, from Tyuratum. The last of the series was launched only a few hours before the 2-yearly launch window ended. This pair was due to arrive in mid-March, 1974, and includes French equipment for studying solar radio emissions. Payload weights are believed to be similar to earlier Mars craft. Russia has used Mariner 9 information in planning landing points; US scientists hope that reciprocal information will help them with their Viking landings in 1975.

Summary of Results Results of data and pictures received from Mars 2 and 3 were reported in

August 1972, when it was stated that their work had been completed. Mars 2 had made 362 revolutions, and Mars 3, in its much more elliptical orbit, 20. The differing orbits had enabled comparative studies to be made of Mars's magnetic field, essential 'to understanding not only the nature of Mars, but the origin and evolution of the solar system'. The spacecraft had travelled more than 621 million miles (1000 million km); 687 communication sessions had taken place, 448 of them while in Martian orbit. Like America's Mariner 9, Mars 2 and 3 were able to study the 3-month dust-storm which began in October 1971, and Soviet scientists concluded that winds were not constant, but most likely occurred only in the initial phase, and that the particles, mostly of silicate composition, took months to fall 'in the absence of supporting vertical currents in the atmosphere'. The dust-clouds were at least 5–6 miles (8–10 km) high; during the storm the surface temperature dropped by 20–30°C, while the atmosphere got warmer as it absorbed solar radiation. When the storm ended surface temperatures taken from the southern hemisphere, where summer was ending, across the Equator to the northern hemisphere, ranged from 13°C to −93°C; at the North Pole cap they dropped to −110°C. The surface cooled very quickly after sunset, indicating a low conductivity, such as dry sand or dust in a rarified atmosphere. The Martian 'seas', or dark areas, were warmer than the 'continents', or light areas, and in some cases the 'seas' took longer to cool, suggesting a rock surface with greater heat conductivity. Atmospheric pressure varied with altitude but at average surface level was about $\frac{1}{200}$th of earth's. Mars 3 measured heights of 1·8 miles (3 km) and depressions of 0·6 miles (1 km) below average level. During the dust-storm the dust-clouds rose as high as 6 miles (10 km). Water-vapour content of Martian atmosphere did not exceed 5 microns of precipitated water—2000 times less than earth's. Immediately above the surface the atmosphere was mainly carbon-dioxide, but at 62 miles (100 km) altitude was broken up by ultra-violet solar radiation into a carbon-monoxide molecule and an oxygen atom. Traces of oxygen were recorded up to altitudes of 430–480 miles (700–800 km), where its concentration was 100 atoms per cu cm. Photographs had revealed interesting twilight phenomena, including an airglow 124 miles (200 km) beyond the terminator, or border line between day and night, and changes in surface colour close to the terminator. The severe climatic conditions did not rule out the possibility of some form of life, though at best it was likely to be no more than micro-organism or plants.

METEOR

Reports by Meteor satellites of Arctic ice conditions, cloud formations in the USSR, and warnings of tropical storms, have saved many lives. Irrigation planning has been improved as a result of data on snow cover in the Tien Shan and Himalaya Mountains, and 10% of sailing time for all Soviet ships is estimated to have been saved because Meteors can provide the ships with the best courses to avoid storms, sea states, winds and ice conditions.

Meteor photographs, by now numbered in hundreds of thousands, have revolutionized Soviet weather-forecasting techniques. In 1 hr the 2 TV cameras aboard each satellite cover 11,583 sq. miles (30,000 sq. km), automatically adjusting their exposure times to match lighting conditions on the daylight side, storing and then transmitting the pictures. On earth's dark side heat emissions and cloud formations are measured by infra-red sensors. Radiometers measure the radiation balance of the earth–atmosphere system; clouds, ice and snowfields reflect about 80% of solar radiation, and measurements of incoming energy are regarded as essential for reliable forecasts. 2 diagrammatic maps are compiled every 24 hrs.

Basically the system provides a daily weather review from more than two-thirds of the globe. Soviet sources say it started to function on April 27, 1967, with the launch of Cosmos 154 and 156. Western observers noted 14 Cosmos meteorological test satellites, beginning with 44 and 45 in September 1964. The second of these returned a film capsule. Subsequent Cosmos, announced as 'metsats', included 144, 156, 184, 206 and 226; their orbits, varying by only 12 secs, were phased to provide almost continuous coverage of the earth's surface. Space 'panoramas' are transmitted to meteorological organizations in Siberia and the Soviet Far East, as well as to several other countries. The satellites are also studying clear air turbulence—strong currents in the upper layers of the atmosphere used by intercontinental jets.

Meteor 1 itself was launched from Plesetsk on March 26, 1969, into a 400 × 443-mile (644 × 713-km) orbit with 81° incl. Launches continue at a rate of 4 a year, so that 2 or 3 are always operational, providing a continuous survey of atmospheric conditions from Pole to Pole in a band up to 932 miles (1500 km) wide. Information must be 'dumped', in passes lasting only a few minutes, into the 3 reception centres of the USSR Hydrometeorological Service in Moscow, Novosibirsk

Meteor Satellite in Moscow Space Pavilion

Meteor reception antenna at Novosibirsk

(Siberia), and Khabarovsk (Pacific Coast); processing the data takes 1½ hrs. Orbital life of Meteors 1–10 is about 60 years; Meteors 11, 12 and 13, however, launched in March, June and July 1972, into 500–600-mile (800–960-km) orbits, have 500-year lives before decay.

Spacecraft Description   No weight has been given; the satellites are believed to be 16 ft (5 m) long; 5 ft (1·5 m) wide. A large cylindrical body carries meteorological instruments at the earth-facing end; a pair of large solar panels, rotated by a drive mechanism so that they remain sun-oriented, charge the electical storage batteries. The instruments are kept pointing towards the earth by a 3-axis attitude control system, which includes reaction wheels.

MOLNIYA

History   Molniya is part of the Soviet Union's communication system, providing TV, telephone and telegraph links to 33 Orbita ground stations in the USSR, plus 1 in Cuba and 1 in Mongolia. It is regularly used to transmit facsimiles of newspaper pages to remote areas. The 24-hr TV service provided to the Far North and Far East territories, said Moscow radio in February 1972, had 'saved the national economy hundreds of millions of roubles' by eliminating the need for a network of ground relay stations. A radio-isotype power supply had already been tested in a Cosmos flight; it was intended to replace Molniya's solar batteries with atomic reactors powerful enough to transmit directly to TV aerials without being boosted by Orbita stations.

Under a US–USSR agreement of September

30, 1971 Molniya 2 satellites are to be used as 'hotline stations', with a complementary service being provided by US Intelsat satellites. A ground installation at Fort Derrick, Maryland, about 45 miles (72 km) from the White House, will link the US President, via Molniya 2s, to the Kremlin. It will have 2 receive/transmit antenna stations with parabolic reflectors of over 40 ft (12 m) dia. An Intelsat ground station is being located near Moscow. This 'hotline' system to remove the danger of accidental war and allow direct conversation between the heads of the Super Powers in times of crisis, will replace the 'terrestrial' cable hotline, routed Washington–London–Copenhagen–Stockholm–Helsinki–Moscow, operated by ITT, with backup route Washington–Tangier–Moscow, operated by RCA Globcom.

Molniya 1   These are placed in highly elliptical orbits from Plesetsk by A2e with approx. 24,800 miles (40,000-km) apogee in the N. Hemisphere, and 300-mile (482-km) perigee in the S. Hemisphere, to provide 8–10 hrs continuous communications through the USSR and associated countries. Orbital changes are made by onboard rocket engines, to synchronize available hours. By the end of 1972 22 Molniya 1s had been launched. The first, L. April 23, 1965 from Tyuratum, provided 'many months' of TV and phone links between Moscow and Vladivostok. Second, L. October 14, 1965 from Tyuratum, provided experimental TV, phone and telegraph communications for 5 months before decaying. Third, L. April 25, 1966 from Tyuratum was used for an exchange of colour TV with France, and for earth pictures showing cloud-patterns on a global scale. Colour photos of earth started in 1967. Molniyas also provide communication links during manned

Molniya 2 at 1973 Paris Air Show. *Far left*, tri-lens camera for earth photos

Soyuz flights. Performance has been steadily upgraded to improve radiation resistance, decrease noise levels and increase power output. The 20th Molniya 1, L. April 4, 1972 from Plesetsk, also carried France's SRET 1 (environmental research satellite) into orbit, under a 1970 collaborative agreement. SRET 1 was released 2 secs later than Molniya. Latest versions of Molniya 1 weigh about 1800 lb (816 kg). Cylindrical in shape with conical top, they are 11·3 ft (3·4 m) long and 5·2 ft (1·6 m) dia., with 6 paddle-wheel solar wings, providing 500–700 W of power. Two 3-ft (0·9-m) dish antennae extend from the base, which also contains an orbital correction engine. 3 transceivers have estimated life of over 40,000 hrs; TV transmissions at 625 lines per frame.

Molniya 2 More advanced and slightly heavier, approx. 2750 lb (1250 kg). 4 launches to end of

1972, by A2e from Plesetsk, in November 1971, and May, September and December 1972, into approx. 300 × 25,000-mile (483 × 40,230-km) orbits. Like recent Molniya 1s, they are believed to be windmill-shaped, with 6 solar wings and 5-year life.

POLYOT

Polyots 1 and 2 (L. November 1, 1963, and April 12, 1964) were 'the first manoeuvrable earth satellites'. With estimated weight of 1322 lb (600 kg), Polyot 1, placed in an initial orbit of 210 × 367 miles (339 × 592 km), was successfully manoeuvred several times, ending in a 212 × 893-mile (343 × 1437-km) orbit with a 25-year life. Polyot 2, which had an 18-day life, made several similar manoeuvres, and also changed its inclination from 58° to 60°. America's first orbital manoeuvres were made by Gemini 3 on March 23, 1965.

PROGNOZ

History Satellite designed specifically to study solar radiation, the physical processes taking place on the sun, and the interference they can cause with space communications.

Prognoz ('Forecast') 1 L. April 14, 1972 by A2e from Tyuratum. Sphere-shaped, with 4 cruciform solar panels. Wt 1890 lb (857 kg). Orbit 600 × 124,000 miles (965 × 200,000 km), gives it a 10-year life. Incl. 65°.

Prognoz ('Forecast') 2 L. Jun 29, 1972 into

Prognoz 2, showing solar radiation
instrument array

341 × 124,000-mile (550 × 200,000-km) orbit,
with 65° incl. In apogee when Prognoz 1 is in
perigee; this provides simultaneous studies of solar
wind from different points of near-earth space,
during their 4-day orbits. Prognoz 2 carries French
equipment to study solar wind, outer regions of
magnetosphere, gamma rays and solar neutrons.

Prognoz 3 L. February 15, 1973. Wt 1863 lb
(845 kg); orbit 367 × 124,000 miles (590 × 200,000
km). Incl. 65°. This was equipped to study solar
gamma-ray and X-ray emissions. Its readings are
synchronized with measurements from earth-
based observatories. The highly eccentric orbit
gives comparative measurements from both near-
earth and from outside the upper atmosphere and
magnetosphere.

PROTON

History A series of 4 satellites to study the nature of cosmic rays of high and super-high energy, and their interaction with the nuclei of atoms. Intercosmos 6, nearly 4 years later, was described as a logical development of the Proton stations.

Proton 1-3 L. July 16, 1965, November 2, 1965 and July 6, 1966, by D1 from Tyuratum. Wt 26,895 lb (12,200 kg), they were of record size at that time. Their orbits were 118 × 391 miles (190 × 630 km), and they had a 3-month life.

Proton 4 L. November 16, 1968, by D1 from Tyuratum. Wt set a new record of 37,500 lb (17,000 kg). Orbit of 307 × 158 miles (493 × 255 km), gave an 8-month life. Experiments included study of the energy spectrum and chemical composition of primary cosmic particles, and the intensity and energy spectrum of gamma rays and electrons of galactic origin.

SPUTNIK

History When Sputnik 1, the world's first artificial satellite, soared into orbit on October 4, 1957 it acted as the starter's pistol in the Soviet–American race to put men on the moon. In America its launching brought bitter disappointment; earlier the same year a US Army plan, supported by Dr Wernher von Braun, who wanted to use his Redstone rocket to put up a satellite in September, was turned down. Had it been successful the United States would have been one month ahead of Russia with man's entry into space. As it was, Russia was to put up Sputnik 2 before America joined the contest with Explorer 1 on January 31, 1958.

The whole series of 10 appears to have been devoted to developing the ability to place men in earth orbit, using at least 6 dogs for the purpose. All were launched from Tyuratum over a period of $3\frac{1}{2}$ years (*see* following pages for details). Cosmos 1 was also Sputnik 11, and for a time the numbers of the 2 series overlapped. Sputniks 19, 20 and 21 were believed to be early attempts at launching Venus probes. Sputniks 22 and 24 were attempts to launch towards Mars; and Sputnik 25 (January 4, 1963) was a Luna failure. From that point, Russia herself allotted Cosmos numbers to any launch failures occurring in non-Cosmos programmes.

Launchers The standard Vostok vehicle ('A'), based on the original Soviet ICBM, providing lift-off thrust of 1,124,000 lb (509,840 kg), was used for Sputniks 1–3; for most of the subsequent launches an upper stage, providing 199,000 lb (90,260 kg) was added to the Vostok vehicle.

Sputnik 1 L. Oct 4, 1957. Wt 184·3 lb (83·6 kg). Orbit 588 × 142 mi (947 × 228 km). Incl. 65°. The world's first artificial satellite. It consisted of a polished metal sphere, dia. 1 ft 11 in. (0·58 m), with 4 whip-type aerials from 4 ft 11 in. (1·5 m), to 9 ft 6 in. (2·9 m) long. Instrumentation included radio telemetry and devices for measuring the density and temperature of the atmosphere and concentrations of electrons in the ionosphere. It transmitted for 21 days; decayed after 1400 orbits in 96 days. **Sputnik 2** L. Nov 3, 1957. Wt 1120 lb (508 kg). Orbit 1038 × 140 mi (1671 × 225 km). Incl. 65°. The world's second satellite. It contained a spherical pressurized container to carry the dog Laika, the first living creature in space, to obtain data on effects of weightlessness on living organisms. Transmissions lasted for 7 days, when Laika presumably died painlessly when her oxygen supply ran out. Spacecraft re-entered and decayed after 2370 orbits in 103 days. **Sputnik 3** L. May 15, 1958. Wt 2926 lb (1327 kg). Orbit 1168 × 135 mi (1880 × 217 km). Incl. 65°. Cone-shaped, 5 ft 8 in. (1·73 m) dia. at base, and 11 ft 7 in. (3·57 m) long, it carried instruments to study the earth's upper atmosphere, solar radiation, etc. Decayed after 690 days. **Sputnik 4** L. May 15, 1960. Wt 10,009 lb (4540 kg). Orbit 229 × 194 mi (368 × 312 km). Incl. 65°. First Sputnik for 2 yrs, with Lunas 1–3 intervening. This was believed to be a test flight for a manned Vostok; recovery failed when the cabin went into higher orbit and finally re-entered over 5 years later on October 15, 1965. **Sputnik 5** L. Aug 19, 1960. Wt 10,141 lb (4600 kg). Orbit 190 × 211 mi (305 × 339 km). Incl. 80°. Second Vostok trial. 2 dogs, Belka and Strelka, were ejected and recovered by parachute after 18 orbits. **Sputnik 6** L. Dec 1, 1960. Wt 10,060 lb (4563 kg). Orbit 116 × 165 mi (186 × 265 km). Incl. 65°. Third Vostok trial; recovery failed, and canine passenger was killed one day later. **Sputnik 7** L. Feb 4, 1961. Wt 14,290 lb (6482 kg). Orbit 203 × 138 mi (327 × 223 km). Incl. 65°. Apparently a test flight in preparation for Sputnik 8, the first attempt at a Venus fly-by. Decayed Feb 25, 1961. **Sputnik 8** L. Feb 12, 1961. Wt 14,275 lb (6474 kg). Orbit 198 × 123 mi (318 × 198 km). Incl. 65°. Launched Venus 1, Russia's first Venus probe, from earth-parking orbit. Sputnik 8 decayed Feb 25, 1961. **Sputnik 9** L. Mar 9, 1961. Wt 10,362 lb (4700 kg). Orbit 114 × 155 mi (183 × 250 km). Incl. 65°. Fourth Vostok trial; dog Chernushka successfully recovered on

Sputnik 1, Man's first satellite

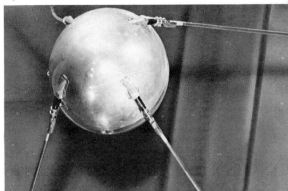

same day. **Sputnik 10** L. Mar 25, 1961. Wt 10,351 lb (4695 kg). Orbit 111 × 153 mi (178 × 246 km). Incl. 65°. Fifth Vostok trial; dog Zvezdochka successfully recovered after 1 orbit. Yuri Gagarin, Russia's first man in space, went into orbit 18 days later.

VENUS

History After more than 10 years of persistent work, and many disappointments in the early years, Soviet scientists have achieved remarkable successes in penetrating the mysteries of Venus. Venus 3 was man's first spacecraft to reach a planet (March 1, 1966); but it plunged into the hostile atmosphere without returning any planetary data. Russia was able to claim the first transmissions from a planet's surface following the successful descent of Venus 7 on December 15, 1970; while there may be some doubt whether this reached the true surface, or fell on mountain peaks, with the result that it tumbled, and abruptly ceased transmissions after 23 mins, there can be no doubt at all that Venus 8 transmitted from the surface for about 1 hr on July 22nd, 1972, before its resistance to the temperatures of up to 500°C and atmospheric pressures up to 100 times those on earth, was finally overcome.

The painstaking persistence of Soviet scientists in using the knowledge gained on each mission to improve both the techniques and spacecraft on each succeeding flight can be followed by a study of the 8 missions detailed below. The information has gradually been gleaned as facts about past missions have been released during the project's progress.

With its clouds reaching an altitude of up to 37 miles (60 km) and surrounding the planet with what has been described as 'a heavy oily smog', Venus presents much greater technical problems than Mars for exploring spacecraft. This may be one explanation for the extent of Soviet interest in Venus, which sometimes seems disproportionate to the likely rewards. By the time it was announced that Venus 8 was on course in March 1972, at least 16 launching attempts had been made. Venus designations are given to spacecraft only when they are safely on course; failures are merely given the next available Cosmos number. Evidence of these failures is to be found in the tables of space launches compiled by Western observers. Russia prefers to operate her planetary probes in pairs, so that transmissions can be compared. This was achieved with Venus 2 and 3, and with Venus 5 and 6; but what should have been Venus 8, accompanying Venus 7 in August 1970, failed to achieve escape velocity, and was written off as Cosmos 359. Similarly, what was clearly intended to be Venus 9, accompanying Venus 8 in March 1972, and launched 4 days later, had to be written off as Cosmos 482 when the escape stage fired only partially, and left it stranded in an elliptical 127 × 6090-mile (205 × 9800-km) orbit with a 6-year life. Venus 9 was intended to land on the planet's dark side, to provide comparative readings with those sent by Venus 8 from the sunlit side.

However, as Soviet scientists developed the use of structural strengthening and ablative materials, and found ways of adjusting the descent rates to achieve greater penetration before their spherical descent craft were destroyed, confidence grew. Following the successful Venus 8 launch, advance information was given that a descent was intended —almost certainly the first time in Soviet space history that flight plan details were published in advance.

Spacecraft Description　Venus spacecraft consist of a cylindrical main section, containing mid-course propulsion engine, telemetry, attitude control, guidance, sensory and power systems with solar panels. The sperical descent module represents about one-third of the usual payload of around 2601 lb (1180 kg). Total weights have not risen much as the series has progressed, as mission details on the following pages reveal; but increasing knowledge has made it possible to use less of the weight on protection from pressure and temperature, and more on scientific content. Centre of gravity is offset to assist self-orienting during descent, with separate hermetically sealed compartments for the instrumentation and dual-parachute descent system. Launch into earth-parking orbits is from Tyuratum by 140-ft (43-m) high Voskhod rockets providing lift-off thrust of 1,124,000 lb (509,840 kg). A 2-engine 2nd stage adds 308,650 lb (140,000 kg) thrust, and there is a 3rd 'escape' stage. In the case of Venus 8 it was announced that this burned for 243 secs, and accelerated the spacecraft to 11·5 km per sec (25,731 mph or 41,433 kph), 'somewhat greater than second cosmic velocity'.

Experience with Venus 4, 5 and 6 showed that the design of the main spacecraft was satisfactory; but the landing vehicle was redesigned to withstand external pressures of up to 180 atmospheres and 530°C, increasing its weight over Venus 5 and 6 by about 220 lb (100 kg). The parachute canopy was made of heat resistant cloth able to withstand temperatures up to 530°C. Published weight figures vary slightly; the total given for Venus 7 was once again 2601 lb (1180 kg).

Western suggestions that Venus 7 ended its life on mountain peaks are not fully accepted by Soviet scientists, some of whom consider that all

Venus 4: landing capsule at bottom

mountains would have been eroded into sand and dust by the hot atmosphere. But a light wind, they concede, could have raised the dust into the atmosphere, and caused dense clouds which would cause echoes in radio frequencies and discrepancies in radio-altimeters. The performance of Venus 7, and the knowledge it added to that gained on earlier missions, made another major redesign possible for the Venus 8 lander.

Since Venus 4–7 had all studied the planet's night side, it was decided that Venus 8 should aim for the sunlit side, to check the amount of light reaching the surface through the cloud layers; this, it was hoped, would indicate how far the solar rays penetrated the thick atmosphere, and perhaps explain how the heating of the atmosphere to such high temperatures took place.

A major advance on Venus 8 was a dual antenna, which enabled it to transmit 60-min data when it succeeded in reaching the sunlit side on July 22, 1972. Additions like this were possible because Venus 7 experience showed that weight-consuming insulation and protection against the elements could be reduced. A refrigeration unit was added to cool the interior below freezing during the early part of the descent; the dual-parachute system, first tried on Venus 7, aided the effectiveness of this by enabling Venus 8 to drop rapidly through the upper atmosphere and then slow down as it approached the surface (*see page* 180). At touch-down a separate antenna, tripod-mounted with large pads and able to withstand heavy winds, was thrown off, coming to rest several feet away. This was to ensure transmission even if the main craft rolled or fell so that its primary antenna was not pointed to earth. In the event, it was possible to use both antennae; the first 13 mins of data came from the primary antenna, and included temperature readings, atmospheric pressure and light levels; 20 mins of data concerning the nature of the hard surface then came from the secondary antenna, with the final 30 mins again from the main antenna.

Note: Figures given for weights and temperatures vary a good deal, even in official statements, and translation difficulties with such technical subjects

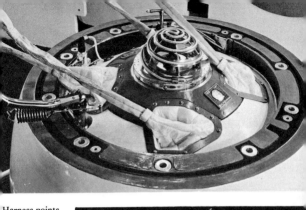

Harness points on Venus 4 descent capsule

Venus 4 descent parachute

also lead to inaccuracies. There is constant confusion between Centigrade and Fahrenheit, which sometimes accounts for inconsistencies in temperatures given on the following pages.

Venus 1 L. Feb. 12, 1961 by Sputnik 8 from Tyuratum. Wt 14,275 lb (6474 kg); Venus wt 1419 lb (643·5 kg). Radio contact lost at 4·7 million mi (7·56 million km), but spacecraft passed 62,000 mi (99,800 km) from planet. This was the first Soviet planetary flight. **Venus 2** L. Nov 12, 1965 by A2e from Tyuratum. Wt. 2123 lb (963 kg). Passed 14,900 mi (23,950 km) from Venus on Feb 27, 1966, but failed to return data. **Venus 3** L. Nov 16, 1965 by A2e from Tyuratum. Wt 2116 lb (960 kg). After a 105-day flight, impacted on Venus on Mar 1, 1966, man's first spacecraft to reach a planet. It failed, however, to return any planetary data. **Venus 4** L. Jun 12, 1967 by A2e from Tyuratum. Wt 2438 lb (1106 kg), of which 844 lb (383 kg) was the entry vehicle. After a 128·4-day flight, presumed to have impacted on Oct 18, 1967, one day before US Mariner 5 fly-by. Descent capsule transmitted data during 94-min. parachute descent. Higher temperature sent back was 540°F; also sent measurements of pressure, density and chemical composition of atmosphere before ceasing transmissions, possibly as a result of striking one of 3 mountain ranges recently detected by radar bounces. This flight enabled later spacecraft to be redesigned so that they could penetrate more deeply and survive longer. **Venus 5** L. Jan 5, 1969 by A2e from Tyuratum. Wt 2491 lb (1130 kg), of which 893 lb (405 kg) was the entry vehicle. After a 131-day flight the descent capsule separated at a distance of 22,992 mi (37,002 km) from Venus. Initial entry velocity of 6·95 mi per sec (11·2 kps), was reduced by atmospheric braking to 470 mph (756 kph) before deployment of main parachute one-third the size of that on Venus 4. The capsule entered the Venusian atmosphere on the planet's dark side on May 16, 1969, and transmitted data for 53 mins, travelling 22·4 mi (36 km), into the atmosphere before being crushed. **Venus 6** L. Jan 10, 1969 by A2e from Tyuratum. Wt 2491 lb (1130 kg). After a 127-day flight, the descent capsule separated at a distance of 15,535 mi (25,000 km), and 24 hrs after Venus 5, entered the Venusian atmosphere on the planet's dark side, and transmitted data for 51 mins, descending 23·5 mi (37·8 km), into the atmosphere before being crushed. *Joint Results* Data from both spacecraft revealed that the concentration of nitrogen with inert gases in the atmosphere of Venus is from 2–5%; oxygen does not exceed 0·4%;

carbon dioxide represents 93–97%. Water-vapour content was very low. During descent, temperature readings ranged from 77°F–608°F. Pressure readings ranged from 7·35 psi (0·5 kg sq cm) to 396·9 psi (28 kg sq cm). Extrapolation of data from both spacecraft suggested surface temperatures on Venus might range from 400–530°F, while the pressure would be from 882–2058 psi (62–144 kg sq cm). Both spacecraft were believed to have been crushed when the ambient pressure exceeded 27 atmospheres—about 400 psi (28 kg sq cm); and the temperatures exceeded 600°F. **Venus 7** L. Aug 17, 1970, by A2e from Tyuratum. Wt 2601 lb (1180 kg). After a 120-day flight, entry into the planet's atmosphere began on December 15. Distance from earth at that time was 41 million mi (60·6 million km). After separation, the descent craft's speed was reduced aerodynamically from about 25,700 mph (41,400 kph), to 450 mph (724 kph). It was subjected to 350G, and temperature differences between the shock wave and vehicle reached 11,000°C. The parachute was opened 37·3 mi (60 km) above the surface, when external pressure was about 0·7 atmospheres. The parachute canopy, of cloth designed to withstand temperatures up to 530°C, was bound by a Kapron cord, so that a fast descent was made through the upper layers; the cord was made to fray and tear apart in the lower layers, which meant that the parachute canopy would then open fully and slow the descent so that the lower layers could be studied more carefully. During this period only the gradually increasing temperatures were transmitted to earth; the signals took 3 mins 22 secs to reach earth. According to a Tass report, changes in descent velocity signals showed that the probe, which carried pennants with a picture of Lenin and the Soviet Union's hammer-and-sickle emblem had landed. Moscow radio said that at this point the radio signals dropped to 1/100 th of the descent strength, probably because the antenna's axis had deviated from the direction of earth after landing. Signals continued for 23 mins after landing; temperatures were 475°C plus or minus 20°C; pressures were 90 atmospheres plus or minus 15. Density of the atmosphere on the surface was about 60 times greater than on the earth's surface.

During the flight of Venus 7 its instruments sent back readings of the powerful solar flare which began on Dec 10, 1970; it was possible to compare Lunokhod 1 readings received simultaneously from the lunar surface, and from satellites and ground observatories. When Venus 7 landed Soviet scientists were able to claim that for the first time they were receiving information simultaneously from 2 celestial bodies. **Venus 8** L. Mar 27, 1972 from Tyuratum. Wt

Venus 8: *right*, high gain antenna; *left centre* low gain antenna

2601 lb (1180 kg). After a 117-day flight, and having travelled 186 million mi (300 million km), Venus 8 reached the planet on Jul 22, when the latter was 67 million mi (108 million km) from earth. 86 communication sessions had taken place during the flight to control the course and check the onboard systems. It was manoeuvred to a touchdown in the narrow, crescent-shaped sunlit portion of the planet visible from earth; the landing site, about 1800 mi (2896 km) from the site of Venus 7, was selected to minimize the earth-distance. Earth was relatively low on the Venusian horizon, so Soviet scientists were able to obtain readings from the sunlit portion; it was early morning at the landing site, with the sun equally low on the local horizon. As the spacecraft entered the upper atmosphere the descent module was separated, while the service module went on to burn up in the atmosphere. The entry speed of about 25,908 mph (41,696 kph) was reduced by aerodynamic braking, and the parachute deployed at 560 mph (900 kph). During descent, a refrigeration system, involving a compressor and heat exchange unit, was switched on to offset the 500°C temperature found on earlier missions. Transmissions of temperatures, pressures, light levels and descent rates were interrupted for 6 mins during descent for calibration of onboard instruments; touchdown was at 09.29 GMT. Soviet scientists promised the details would be issued when information obtained had been processed, and said that so much heat was stored in the dense and heavy atmosphere that night and day temperatures levelled out, even though each lasted about 2 terrestrial months. Surface light levels were considerably lower than earth's. Surface temperatures probably varied only 1°F; they were such that tin and lead would melt, and iodine, mercury, bromine and sulphur would evaporate.

Planet Venus

Of the 9 planets, it is the second nearest to the sun, 67 million miles (108 million km), compared with earth's 93 million miles (150 million km). Diameter is 7700 miles (12,400 km); earth's 8000 miles (12,900 km). Like Mercury and Pluto, Venus has no known satellites. It is often so bright in a clear night sky that when appearing just above the horizon, it has frequently started 'flying saucer' scares in England. This is the result of sunlight reflecting from the heavy cloud layers, probably 40 miles (64 km) thick. Another effect is that on the surface, even in the middle of the Venusian day (which lasts 245 earth days), it is no brighter than the dullest winter day on earth. When Venus is closest, only the dark side faces earth. Soviet scientists consider that it is going through a stormy process of geological evolution, and by studying it in detail they hope to discover the phenomena that take place more slowly on earth. The study may well reveal what awaits our own planet in the future.

Launch opportunities for spacecraft occur only

once in every 19 months; favourable future periods are: October–November 1973; May–June 1975; January 1977; August 1978 and March–April 1980.

ZOND

History When the series began in 1964, the spacecraft were described as 'automatic interplanetary stations and deep spaceflight technology development tests'. For Western observers, this series has been the most puzzling of all Soviet programmes. President Keldysh, of the Soviet Academy of Sciences, confirmed in January 1969, following Russia's first manned docking with Soyuz 4 and 5, that Zond spacecraft were 'adapted for manned flight'. However, he added a warning that such flights should not be expected 'in the next 2 or 3 weeks'. At that time, it seems, Russia was still undecided upon the division of space exploration between manned vehicles and automatic apparatus. But Zond 5 was the world's first spacecraft to make a circumlunar flight and return safely to earth in September 1968; 2 months later, the feat was repeated by Zond 6. Tortoises, insects and seeds carried on board, showed no ill effects. Spurred on by the fact that the Russians were apparently making final circumlunar tests with a manned spacecraft, America took a chance and sent Apollo 8 round the moon at Christmas, 1968; though they did not land, the first manned flight *around* the moon was an achievement ranking second only to the first landing. Perhaps one day we shall learn why the Russians, so conscious of the prestige value of space 'firsts', failed to achieve this one when they clearly had the capability. Probably their rockets did not have sufficient lifting-power for a fully manned Zond. One theory is that they could have sent a single cosmonaut, but felt it was too risky to send one man alone on the first circumlunar flight; if he became ill, or technical problems led to survival depending on manual control of the spacecraft, the workload might well have proved too much for one man.

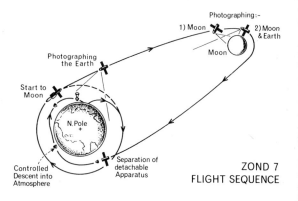

ZOND 7
FLIGHT SEQUENCE

These may be the reasons why details of Zond now appear in *Unmanned Spacecraft*. But it is still possible that the entry will ultimately have to be transferred to *Manned Spaceflight*. Zond 8's activities suggest that, having missed the first circumlunar flight, this spacecraft may well be undergoing tests for the first manned fly-by of Mars, perhaps towards the end of this decade.

Cosmos 21, on November 11, 1963; Cosmos 27, on March 27, 1964; and Cosmos 379, 382, 398 and 434 in 1970 and 1971, were probably unsuccessful Zond missions.

Zond 1 L. Apr 2, 1964 by A2e from Tyuratum. Wt 2094 lb (950 kg). Believed to be a Venus probe, but communications were lost. It finally passed Venus at a distance of 62,000 mi (99,779 km), and went into solar orbit, after course corrections at 350,000 mi (563,270 km), and 8·7 million mi (14 million km) from earth. **Zond 2** L. Nov 30, 1964, by A2e from Tyuratum. Wt 2094 lb (950 kg). Passed within less than 1000 mi (1500 km) of Mars, but failed to return any data, and went into solar orbit. **Zond 3** L. Jul 18, 1965, by A2e from Tyuratum. Wt 2094 lb (950 kg). Took 25 photographs of lunar far side at distances of 5730–7190 mi (9219–11,568 km) and transmitted them to earth 9 days later from distance of 1,367,000 mi (2,200,000 km). Transmission was repeated from 19·5 million mi (31·5 million km) presumably in a test of photographic systems for Venus 2 and 3 later that year. **Zond 4** L. Mar 2, 1968 by D1e from Tyuratum. Wt ?5510 lb (2500 kg). Flight tested new systems 'in distant regions of circumterrestrial space'. **Zond 5** L. Sep 15, 1968 by D1e from Tyuratum. Wt ?5510 lb (2500 kg). First spacecraft to circumnavigate the moon and return to earth; fired out of earth-parking orbit before completion of first orbit into a free-return translunar trajectory. A Soyuz-like vehicle, it

Lunar far side taken by Zond 3. Dark Spot (*right*) is Eastern Sea

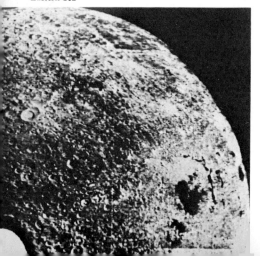

Zond 5 on Soviet ship after sea recovery

consisted of a heat-shielded re-entry module, with camera, scientific package, radio and telemetry, re-entry control and parachute systems, and an instrument compartment with both major course correction and low-thrust vernier engines, and solar-powered batteries. Tortoises, insects, plants and seeds were carried. After passing round the moon with the closest point 1212 mi (1950 km) away, it took earth photos at a distance of 55,923 mi (90,000 km) but no moon photos were mentioned at the time. During the ballistic trajectory return to earth a tape-recorded Russian voice was heard calling out simulated instrument readings. Parachutes were deployed at 4·3 m (7 km); splash-down, at the end of a 7-day flight, was in the Indian Ocean, Russia's first sea recovery. The capsule was taken to Bombay, and flown back from there in a Soviet air-craft. **Zond 6** L. Nov 10, 1968 by D1e from Tyuratum. Wt 6000 lb (2720 kg). Second circumlunar flight, described by Tass as intended 'to perfect the automatic functioning of a manned spaceship that will be sent to the moon'. As it passed around the moon at a minimum distance of 1503 mi (2420 km), the far side was filmed. Re-entry was made by a skip-glide technique; the spacecraft's aerodynamic life was used to bounce it off the atmosphere, at a shallow angle, thus reducing speed and gravitational forces. Recovery, after another 7-day flight, was on land inside the Soviet Union. **Zond 7** L. Aug 7, 1969 by D1e from Tyuratum. Wt 6000 lb (2720 kg). Third cir-cumlunar flight. Far side was photographed from distance of 1243 mi (2000 km); colour pictures of both earth and moon were brought back, following re-entry, again by skip-glide technique, with recovery on Feb 14. **Zond 8** L. Oct 20, 1970 by D1e from Tyuratum. Wt ?8818 lb (4000 kg). Fourth circumlunar flight, with more colour pictures of earth and moon. The spacecraft itself was photographed by an optical telescope at distances up to 173,000 mi (277,000 km) from earth; the telescope was pointed with the aid of an onboard laser. The main variation in this flight was re-entry from the N. Hemisphere, as opposed to the normal S. Hemisphere ap-proach for touchdown on Soviet territory. The result was Russia's second sea-recovery: splashdown was on Oct 27 in the Indian Ocean.

INDEX

188

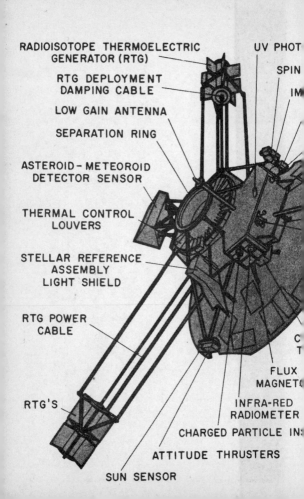

RADIOISOTOPE THERMOELECTRIC
GENERATOR (RTG)

RTG DEPLOYMENT
DAMPING CABLE

LOW GAIN ANTENNA

SEPARATION RING

ASTEROID-METEOROID
DETECTOR SENSOR

THERMAL CONTROL
LOUVERS

STELLAR REFERENCE
ASSEMBLY
LIGHT SHIELD

RTG POWER
CABLE

RTG'S

UV PHOT

SPIN

IM

C
T

FLUX
MAGNETO

INFRA-RED
RADIOMETER

CHARGED PARTICLE INS

ATTITUDE THRUSTERS

SUN SENSOR